Huaming Wu

Motion Compensation for Near-Range Synthetic Aperture Radar Applications

Karlsruher Forschungsberichte
aus dem Institut für Hochfrequenztechnik und Elektronik

Herausgeber: Prof. Dr.-Ing. Thomas Zwick

Band 67

Motion Compensation for Near-Range Synthetic Aperture Radar Applications

by
Huaming Wu

Dissertation, Karlsruher Institut für Technologie
Fakultät für Elektrotechnik und Informationstechnik, 2012

Impressum

Karlsruher Institut für Technologie (KIT)
KIT Scientific Publishing
Straße am Forum 2
D-76131 Karlsruhe
www.ksp.kit.edu

KIT – Universität des Landes Baden-Württemberg und
nationales Forschungszentrum in der Helmholtz-Gemeinschaft

KIT Scientific Publishing 2012
Print on Demand

ISSN 1868-4696
ISBN 978-3-86644-906-0

Editor's Preface

For nearly 100 years now, the use of radar systems has been increasing in our modern society. From industries to daily life, radar systems, originally developed for air and military surveillance, can nowadays be found in weather radars, automatic door opening systems, crash avoidance systems in automobiles, etc. The application possibilities of radar sensor systems in general are not only determined by their accuracy of measurement and resolution, but also their reliability, robustness, size and cost. Compared to other sensor systems such as optical sensors, radar systems using electromagnetic waves have the advantage of working well even under harsh conditions i.e. smoke, haze, night time. The tradeoff is that these systems are often large in size, expensive or do not provide good angular resolution.

Radar scanning is normally achieved mechanically or electronically by making use of the real antenna aperture. By taking successive measurements of a radar, which moves along a line perpendicular to its main radiation direction, a good angular resolution of the radar image can be acquired as well. This method is called synthetic aperture radar (SAR). SAR systems in satellites or other flying platforms have been in operation since the 1950s. The research on near-range SAR systems on the other hand started only in the 1990s, at first for landmine detection and recently as well for automotive applications. The key for a good radar image quality is the compensation of the antenna trajectory deviation from a straight line. The algorithms used therefore in airborne systems cannot be applied to near-range radar systems. This is where the doctoral thesis of Dr. Huaming Wu applies. He has developed a new algorithm for motion compensation especially suited for near range SAR and thoroughly investigated the influence of motion errors on the SAR image. Mr. Wu successfully demonstrated that good near-range SAR imaging is possible with low cost inertial navigation systems.

The presented work of Mr. Wu is a good basis for future near-range synthetic aperture radar. The results will draw attention worldwide and entail more relevant research. I wish him all the best for his future and I am sure that he will produce many more essential contributions in the related field.

Prof. Dr.-Ing. Thomas Zwick
– Director of the IHE –

Forschungsberichte aus dem
Institut für Höchstfrequenztechnik und Elektronik (IHE)
der Universität Karlsruhe (TH) (ISSN 0942-2935)

Herausgeber: Prof. Dr.-Ing. Dr. h.c. Dr.-Ing. E.h. mult. Werner Wiesbeck

Forschungsberichte aus dem
Institut für Höchstfrequenztechnik und Elektronik (IHE)
der Universität Karlsruhe (TH) (ISSN 0942-2935)

Forschungsberichte aus dem
Institut für Höchstfrequenztechnik und Elektronik (IHE)
der Universität Karlsruhe (TH) (ISSN 0942-2935)

Forschungsberichte aus dem
Institut für Höchstfrequenztechnik und Elektronik (IHE)
der Universität Karlsruhe (TH) (ISSN 0942-2935)

Forschungsberichte aus dem
Institut für Höchstfrequenztechnik und Elektronik (IHE)
der Universität Karlsruhe (TH) (ISSN 0942-2935)

Fortführung als
"Karlsruher Forschungsberichte aus dem Institut für Hochfrequenztechnik und
Elektronik" bei KIT Scientific Publishing
(ISSN 1868-4696)

Karlsruher Forschungsberichte aus dem
Institut für Hochfrequenztechnik und Elektronik
(ISSN 1868-4696)

Herausgeber: Prof. Dr.-Ing. Thomas Zwick

Die Bände sind unter www.ksp.kit.edu als PDF frei verfügbar oder als Druckausgabe bestellbar.

Karlsruher Forschungsberichte aus dem
Institut für Hochfrequenztechnik und Elektronik
(ISSN 1868-4696)

Band 63   Janson, Małgorzata
          Hybride Funkkanalmodellierung für ultrabreitbandige MIMO-Systeme
          (2011)
          ISBN 978-3-86644-639-7

Band 64   Pauli, Mario
          Dekontaminierung verseuchter Böden durch Mikrowellenheizung
          (2011)
          ISBN 978-3-86644-696-0

Band 65   Kayser, Thorsten
          Feldtheoretische Modellierung der Materialprozessierung mit
          Mikrowellen im Durchlaufbetrieb (2011)
          ISBN 978-3-86644-719-6

Band 66   Sturm, Christian Andreas
          Gemeinsame Realisierung von Radar-Sensorik und
          Funkkommunikation mit OFDM-Signalen (2012)
          ISBN 978-3-86644-879-7

Band 67   Wu, Huaming
          Motion Compensation for Near-Range Synthetic Aperture Radar
          Applications (2012)
          ISBN 978-3-86644-906-0

Motion Compensation for Near-Range Synthetic Aperture Radar Applications

Zur Erlangung des akademischen Grades eines

DOKTOR-INGENIEURS

der Fakultät für
Elektrotechnik und Informationstechnik,
am Karlsruher Institut für Technologie (KIT)

genehmigte

DISSERTATION

von

M.Sc. Huaming Wu

aus Xiamen, China

Tag der mündlichen Prüfung: 23.07.2012
Hauptreferent: Prof. Dr.-Ing. Thomas Zwick
Korreferent: Prof. Dr.-Ing. Gert F. Trommer

Preface

This dissertation was written during my work as a PhD candidate at the Institut für Hochfrequenztechnik und Elektronik (IHE) at the Karlsruher Instituts für Technologie (KIT).

First and foremost, I wish to express my deep sense of gratitude to my supervisor Professor Thomas Zwick. He has offered me generous and extremely valuable guidance and support throughout the whole process. He has kept me focused and I will be forever grateful. I would like to thank Professor Trommer, director of the Institut für Theoretische Elektrotechnik und Systemoptimierung (ITE), for valuable advice and for being my second supervisor. I would also like to thank Professor Werner Wiesbeck for offering me the position in IHE in the first place.

Additionally, I would like to thank all my colleagues in IHE who have made the years here an important and fun part of my life. In particular, I would like to thank Dr. Jung-hyo Kim for the very helpful discussions in my first years of SAR research; I would like to thank Xuyang Li, Lukasz Zwirello and Lars Reichardt for their help during the SAR measurements; I would like to thank Leen Sit for proofreading this thesis. I would also like to thank Justus Seibold and Dr. Armin Teltschik from ITE for their helpful suggestions about inertial measurements.

I would also like to convey thanks to the Siemens/DAAD Post Graduate Program and Karlsruhe House of Young Scientists (KHYS) for providing the financial support.

Finally, none of this would have been possible without the unwavering support of my parents - Meihua Zhou and Zhangyu Wu. To my wife - Jiamin Feng, I know I can always count on you and for that I feel really lucky.

Karlsruhe, in May 2012
Huaming Wu

Table of Contents

4 Range Motion Compensation Algorithm — 87

5 SAR Demonstrator — 107

6 Measurements — 113

Acronyms and Symbols

Acronyms

2D	Two-Dimensional
3D	Three-Dimensional
ACC	Adaptive Cruise Control
AD	Analog Devices
ADC	Analog-to-Digital Converter
AM	Amplitude Modulation
AO	Analog Output
ARW	Angle Random Walk
BSD	Blind Spot Detection
CAN	Controller Area Network
CSA	Chirp Scaling Algorithm
DAQ	Data AcQuisition
DBF	Digital BeamForming
DCS	Data Collection Surface
DEM	Digital Elevation Model
DOA	Direction Of Arrival
DoD	Degree of Discomfort
DoF	Degree of Freedom
FFT	Fast Fourier Transform

FNR	Furthest range to Nearest range Ratio
FoV	Field of View
GPS	Global Positioning System
HPBW	Half Power BeamWidth
IDP	Image Display Plane
IFFT	Inverse Fast Fourier Transform
IF	Intermediate Frequency
IMU	Inertial Measurement Unit
INS	Inertial Navigation System
IRF	Impulse Response Function
IRPR	Integrated Resolution Portion Ratio
ISLR	Integrated SideLobe Ratio
ITE	Institut für Theoretische Elektrotechnik und Systemoptimierung
KIT	Karlsruher Institut für Technologie
LOS	Line-Of-Sight
MAV	Micro Aerial Vehicle
MEMS	Micro Electronic Mechanical System
MSR	Motion frequency to Synthetic aperture frequency Ratio
POSP	Principle Of Stationary Phase
PRF	Pulse Repetition Frequency
QPE	Quadratic Phase Error
RCMC	Range Cell Migration Correction
RCM	Range Cell Migration
RDA	Range Doppler Algorithm
RMA	Range Migration Algorithm
rms	root mean square

RPM	Revolution Per Minute
RVP	Residual Video Phase
SAR	Synthetic Aperture Radar
SLL	SideLobe Level
UAV	Unmanned Aerial Vehicle
VCO	Voltage Controlled Oscillator

Capital Letters

A	constant amplitude of signal
B	bandwidth
$\mathbf{C}$	direction cosine matrix
D	maximum oveall dimension of antenna
$F_{\Delta r_{\mathrm{azi}}}$	broadening factor of Δr_{azi}
F_{sys}	noise figure of the radar system
G	antenna gain
H_{azi}	matched filter for azimuth compression in range Doppler algorithm
K_a	the parameter of cubic phase errors in the spatial frequency domain caused by Δa_x
K_v	the parameter of quadratic phase errors in the spatial frequency domain caused by Δv_x
L	synthetic aperture interval
N_{acc}	noise density of accelerometer
N_{gyr}	noise density of gyroscope
N_{se}	samples number of each segment in 1-DoF motion measurement algorithm
N_{t}	number of IMU samples within one synthetic aperture length of target t
N_{Taylor}	number of coefficients of Taylor window
P_{95}	the value of 95th percentile

P_r	received power of radar
P_t	transmitted power of radar
D	maximum oveall dimension of antenna
S_{IF}	radar signal in two-dimensional spatial frequency domain
T_{L_t}	synthetic aperture time of a certain target
T_p	duration of chirp signal
V_{Full}	full voltage range of ADC
W_a	envelope of azimuth frequency
W_d	frequency weighting factor of translation vibration in the x and y axes
W_e	frequency weighting factor of rotation vibration
W_k	frequency weighting factor of translation vibration in the z axis

Greek Symbols

$\alpha_{os,azi}$	azimuth oversampling factor
Δr	resolution
$\Delta v, \Delta a$	constant velocity error and constant acceleration error
$\Delta x, \Delta y, \Delta z$	motion errors with respect to the nominal trajectory along x, y and z axes
γ	chirp rate
λ_c	wavelength of center frequency
Φ	elevation half power beamwidth (HPBW)
ϕ	phase variable or elevation angle in antenna beam pattern or roll angle
Ψ	azimuth half power beamwidth (HPBW)
ψ	the angle between the instantaneous slant range vector and zero Doppler plane
σ_{acc}	standard deviation of accelerometer's noise
σ_{accd}	standard deviation of the digitized acceleration
σ_{AD}	standard deviation of quantization noise introduced after digitization

σ_t	target's radar cross-section
θ	pitch angle

List of Constants

c	speed of light in vacuum $2.997925 \times 10^8\,\text{m/s}$
g	standard gravity $9.80665\,\text{m/s}^2$
π	Pi $3.1415...$

Mathematical Notations and Symbols

cos	cosine
exp	exponential
$\mathscr{F}$	Fourier transform
j	imaginary unit $j = \sqrt{-1}$
$\mathbf{a}$	matrix
mod	modulus after division
rect	rectangular window function
sin	sine
tan	tangent

Superscripts and Subscripts

'	relates quantity with regard to short-term trajectory
a	relates quantity to antenna
acc	relates quantity to accelerometer
azi	relates quantity to azimuth
azicom	azimuth compressed
azishift	denotes shift in azimuth
b	relates quantity to body frame

bias   denotes short-term constant bias of IMU

cf   relates quantity to curve fitting

Dop   Doppler

e   relates quantity to Earth-fixed frame

gyr   relates quantity to gyroscope

ideal   relates quantity to the case of ideal rectangular support band

in   denotes space-invariant motion errors

MEFree   relates quantity to motion error-free SAR measurement

ME   relates quantity to motion error-deteriorated SAR measurement

MoCo   relates quantity to motion error-compensated SAR measurement

nsg   non-stop-and-go compensated

os   oversampling

ran-azi   denotes range-azimuth coupling

ran   relates quantity to range

ref   reference

rtaf   range-time azimuth-frequency

rtat   range-time azimuth-time

SC   relates quantity to scene

se   relates quantity to segment in 1-DoF motion measurement algorithm

S   denotes Stolt mapping

ss   relates quantity to sub-swath

t   relates quantity to target or transmitting

var   denotes space-variant motion errors

Lower Case Letters

a acceleration variable

a_v vibration total value

x

$a_{\mathrm{W}x}$	frequency-weighted rms acceleration
d	length from the center of IRF to the first null
f	frequency variable
f_{c}	center frequency of transmitted chirp
f_{m}	frequency of motion error
f_{min}	minimum frequency of transmitted chirp
f_{r}	demodulated range frequency
f_{s}	sampling frequency of IF signal
f_{syn}	synthetic aperture frequency
k_{ant}	broadening factor of azimuth resolution associated with antenna pattern
k_{f}	number of the mimimum periods of the vibration to be quadratic curve fitted
k_{os}	oversampling factor
k_{v}	multiplying factor for vibration
k_{win}	broadening factor of resolution associated with weighting function
$k_x,\ k_y,\ k_r$	spatial frequency of x, y and r
m	index of chirps
p_{a}	the inverse Fourier transform of W_{a}
p_{r}	the inverse Fourier transform of w_{r}
r	slant range
$r_{\mathrm{s,max}}$	the maximum slant range of the furthest target of interest
$r_{\mathrm{t,0}}$	range of the closest approach
s_{acc}	accelerometer's sensitivity
s_{gyr}	gyroscope's sensitivity
s_{ref}	reference function for de-chirping
$s_{\mathrm{t}},\ s_{\mathrm{r}},\ s_{\mathrm{IF}}$	transmitted chirp, received chirp, IF signal
t	time variable

$\hat{t}$	time within one chirp
v	velocity variable
t_d	time delay between the transmitted and received chirp
w_a	azimuth envelope of two-dimensional (2D) raw radar data
w_r	envelope of the transmitted chirp or range frequency
x_am	antenna's nominal coordinate x at the start of m_th chirp
x, y, z	cartesian coordinates

1 Introduction

Radar-based sensors have advantages over optical and ultrasound systems in terms of the all weather capability and the ability to operate through smoke and at night. Owing to the development of hardware and consequently the fall of prices, the first step of the democratization of radar is clearly under way. One of the promising civil applications of radar is automotive radar, whose market has been expanding at about 40 percent a year [1]. Recently more and more cars are equipped with radar sensors for various applications such as adaptive cruise control (ACC), blind spot detection (BSD), collision warning, etc. [2, 3]. Also, radar sensors have been equipped on micro air vehicles (MAV) and unmanned aerial vehicles (UAV) as an altimeter [4, 5] or for proximity warning [6].

It is well known that a good angular resolution of an antenna is achieved through a narrow antenna beam and consequently a narrow field of view (FoV). To address the contradiction between a high angular resolution and a wide FoV of radar sensor, especially in automotive applications, several techniques besides mechanically scanning antenna have been applied. For long range applications, a narrow antenna beam is shaped by using dielectric lens, and multiple beams are generated to cover the whole FoV as well as to achieve a better direction of arrival (DOA) estimation with monopulse technique [7]. As for short range applications, digital beamforming (DBF) radar systems have been developed [8, 9, 10, 11], which digitally scan a wide FoV with virtual narrow beams. However, the practical implementation with regard to routing and isolation between channels as well as computational efforts proves difficult [12]. Furthermore, the angular resolution of DBF or phased array radar system is limited in the range of the half power beamwidth (HPBW) of the antenna array. To achieve an angular resolution beyond this limitation, some "super-resolution" DOA algorithms have been developed [13]. Both the maximum likelihood algorithm [14, 15] and the subspace based algorithms [16, 10] have been proposed for automotive radar network; the former algorithm can provide an optimum solution while the latter algorithms are more computationally feasible. Nevertheless, all of them rely on the fact that signals arriving from different targets are incoherent and the number of targets is known, and therefore the practical use is limited [17].

On the contrary, a synthetic aperture radar (SAR) uses an antenna with wide azimuth beam pattern to virtually form a wide synthetic aperture by traveling along a certain trajectory. The SAR collects and coherently processes the reflected echoes, thereby generating a two-dimensional (2D) image with very high azimuth resolution, which is independent of range and proportional to the real antenna beamwidth [18, 19]. In the last decade, high-performance, compact, low-cost customized SAR systems have been developed and equipped successfully on UAV for remote sensing, surveillance, reconnaissance and environmental monitoring applications [20, 21, 22]. In principle, by combining with SAR techniques, current commercial short range radar products can achieve very high azimuth resolution, and therefore additional applications are expected.

1.1 Near-range SAR Systems

Near-range SAR systems were first developed as the substitute of the airborne SAR system for the preliminary and experimental SAR studies. Compared to the remote sensing airborne SAR systems, such SAR systems were usually equipped on a car and used to image a scenario with a range swath of hundreds of meters long [23, 24].

There are also SAR systems being developed for the specific near-range automotive applications which exploit their high azimuth resolution. For instance, in [25] a short range automotive 24 GHz radar sensor has been used for SAR imaging for parking lot detection, which is usually realized by additional ultrasound sensors [26, 27, 28, 29] or a camera [30, 31]; in [32] a near-range SAR system has been developed for the detection of fawns during pasture mowing.

In addition, the possibility to use SAR systems for landmine detection has been discussed since 1990s [33, 34, 35, 36, 37, 38]. However, there is still no practical landmine detecting automotive SAR system working at stripmap mode, which is primarily due to the lack of approach to address the motion errors arising from the realistic driving environment.

Self-contained navigation system has been of particular interest for MAV applications in the cases where no GPS signals or external cameras are available. By using SAR technique, the onboard radar sensor is able to generate a high resolution image of the surrounding, which can be used to assist the current optical [39, 40] or infrared camera [41] based systems.

1.2 Previous Works and Motivation

A number of algorithms for focusing the raw SAR data have been developed since its debut in 1950s. Most of them were originally developed for remote sensing applications. Therefore some approximations and assumptions due to the restrictions of the hardware, e.g. antenna beamwidth, squint angle, the computation ability of signal processor, have been made when focusing the raw SAR data. The most accurate SAR algorithm is the time-domain correlation algorithm [42], which can be used to process the SAR data acquired with arbitrary beam-width and space sampling trajectory. Neither restriction on the antenna's beamwidth nor motion compensation need to be applied. However, it requires a very high computational effort, which makes it impractical for any true real-time SAR application given the performance of a realistic hardware environment [43, 44]. By restricting the space sampling interval to uniform, frequency-domain algorithms, which perform the focusing in frequency domain with the power of FFT, can substantially reduce the computational burden.

The range Doppler algorithm (RDA), as the most basic frequency-domain algorithm, has been adopted in the cases of stable and high-altitude flying platforms [45]. The chirp scaling algorithm (CSA) can achieve equal or superior performance compared with RDA while avoiding interpolation for range cell migration correction (RCMC) [46]. A modified version of RDA in terms of the realization of RCMC has been proposed in [47, 48, 49], which uses chirp-z transform to perform RCMC therefore also avoids the computationally inefficient interpolation. However, both RDA (including its modified versions) and CSA require restriction on the antenna's beamwidth and the squint angle, which results from the approximation made on the range equation of SAR. A general algorithm has been presented in [50], which considers neither only the first-order term (as RDA) nor the first- and second-order terms (as CSA) of the Taylor series of the signal trajectory, but arbitrary first n-orders terms. Nevertheless, the most accurate frequency-domain algorithm is the range migration algorithm (RMA) [18], which is also called wavenumber domain algorithm [42, 51] or $\omega - k$ algorithm [19]). RMA makes no approximation on the range equation therefore can apply the most exact SAR processing for all squint angles and antenna beamwidths. However, all the frequency-domain SAR algorithms assume that the support band in spatial frequency domain is rectangular or select a rectangular support band (or quasi-rectangular support band as described in [52]) to generate the final SAR image, which results in a great waste of Doppler bandwidth when an ultra-wide beamwidth antenna is used as in the case of automotive short range radar.

There have been a number of publications discussing the influences of motion errors for airborne SAR systems [53, 54, 55, 56, 57, 48, 58, 59, 60], where the motion parameters used for investigations were chosen according to the dynamics of the airplane and the geometry of the SAR operation was also specific for the remote sensing applications. However, very few investigations about the influences of motion errors on near-range SAR applications have been proposed. References with regard to near-range SAR applications either performed the same motion compensation as in the remote sensing applications [61, 62] or assumed that there is no motion error in their applications [25, 35]. For instance, although a project called HOPE had been carried out in order to address the problem arising from the realistic driving environment [63, 37, 64], to date all the proposed landmine detecting SAR systems are still all trailborne systems. Researchers focus on the post processing of SAR imagery by assuming a motion-error-free data set is available. Furthermore, although there is an analytical deduction of influences of phase errors in [18], the relationship between motion errors and phase errors for a SAR system with wide beamwidth antennas is a neglected area.

There are two aspects of motion compensation for SAR applications: obtaining motion errors and compensating motion errors. Motion compensation algorithms can be classified into two groups in terms of the approach to obtain motion errors, i.e. one is to extract them from the raw SAR signal and the other one is to measure them by hardware, such as inertial measurement unit (IMU) and global position system (GPS). Motion compensation algorithms using the former method, termed "autofocus", generally require the imaged scene containing strong point targets [65, 66, 18, 67, 68, 69, 70, 71, 72], which, however, do not necessarily exist in near-range applications. On the contrary, motion compensation algorithms measuring motion data by hardware have no such requirement and have been used since the very beginning of remote sensing SAR applications [53, 54, 73, 74, 22]. In addition, hardware-based motion measuring methods can cope with motion errors with higher frequency than the signal-based motion extracting method [18, 75], and can also work with autofocus algorithms as a trade-off between the complexity of the navigation algorithm and the quality of the SAR image [76, 77].

Thanks to the development of micro electronic mechanical system (MEMS) technology [78, 79], strapdown inertial navigation system (INS) has become more and more affordable for civil applications. For example, INS consists of at least 6 degrees of freedom (DoF): three (triaxial) accelerometers and gyroscopes respectively, combined with GPS has been widely used in home-made MAV and UAV [80]. However, it is not yet possible to build MEMS based INS which gives sub-meter position accuracy over one minute of operation without the aid of other sensors, such as onboard cameras [39, 81], a Vicon camera-based motion tracking system [82, 83], or GPS [84, 85, 86, 87, 88, 89]. In addition, low-cost MEMS based IMUs can be used for navigation only with constraints on the motion of specific applications, such as pedestrian navigation applications [90, 91, 78] or land vehicle navigation applications [92, 93].

In general, azimuth motion compensation is to guarantee uniformly spaced sampling along the azimuth direction, which can be accomplished by linking the pulse emission to a measure of the along-track velocity [61, 94] or by azimuth resampling applied later to the unevenly sampled raw SAR data [42]. Moreover, in [95], a method for compensating the instantaneous azimuth velocity has been presented, where a Doppler compensation phase term has been built as a function of the instantaneous azimuth velocity, without resampling the raw SAR data under certain circumstances.

The motion compensation with regard to space-invariant motion errors, which are with respect to the nominal trajectory as in a stripmap SAR or the center of the scene as in a spotlight SAR, is referred to as first-order motion compensation. It is performed by multiplying the raw SAR data with a common compensation phase term which is built based on the motion data. Furthermore, in order to cope with the space-variant motion errors remaining in the signal history of targets which are away from the reference trajectory or the scene center, several second-order motion compensation approaches have been proposed.

To cope with imaging a wide scene in spotlight SAR applications, the motion compensation algorithm referred to as sub-area algorithm is used, which divides the whole scene into several small areas and then performs motion compensation with regard to the centers of each sub-area, in the end adds each sub-area independently to form the final SAR image of the full scene [96, 74].

To address the problems caused by the high dynamics of airplane for remote sensing SAR applications working in stripmap mode, normal SAR algorithms have been modified. A second-order motion compensation for range-variant motion errors has been added at the stage of the SAR processing after RCMC but before azimuth compression, such as extended CSA [97, 98], extended RDA [99] and extended RMA [100, 101]. Furthermore, at least four types of motion compensation algorithms have been proposed to compensate the azimuth-variant motion errors which cannot be ignored in the case of SAR system with wide beamwidth antenna.

1. Precise topography- and aperture-dependent algorithm [42, 102]. It implements the second-order motion compensation by filtering the signal in the spatial frequency domain with a spatially varying filter for each pixel of the scene. However, the computational burden is so high that it is impractical for real time civil application given the performance of realistic hardware environment.

2. Spot-like motion compensation [103, 52, 58]. It is derived from the aforementioned spotlight SAR case. The stripmap SAR data are divided into several sub-patch along the azimuth direction, and then the spotlight-like motion compensation, which compensates motion errors relative to the sub-patch center instead of the broadside target, is performed on each sub-patch. Though most of the residual motion errors are compensated for the targets at the edge of the sub-patch, the efficiency of such a motion compensation algorithm is inversely proportional to the overlaps along azimuth.

3. Sub-aperture topography- and aperture-dependent algorithm [104, 105]. It performs short time Fourier transform (sub-aperture) along the azimuth direction in blocks, then implements motion compensation for azimuth-variant motion errors referencing to the center of each block before azimuth compression, and finally transforms the SAR data back to azimuth time domain by inverse Fourier transform for the following normal SAR processing. However, a compromise between angle accommodation and trajectory deviation accommodation has to be made by choosing the block size.

4. Frequency division algorithm [106, 107, 108]. It takes a similar approach as the third one but makes the blocks (looks) in the azimuth frequency domain instead

of in the azimuth time domain. As result, it can accommodate azimuth-variant motion errors accurately. However, its computational burden is much higher than that of the third one since motion compensation requiring two full size Fourier transforms has to be performed for each look angle.

From the previous discussion, it can be seen that some critical aspects of motion compensation for near-range SAR applications have not yet been widely explored, which include the SAR algorithm's adaptability to ultra-wide antenna beamwidth, influences of motion errors in the specific near-range field, and the possibility to exploit the geometry of near-range SAR in terms of motion measurement and compensation.

1.3 Goal and Thesis Outline

The purpose of this work is to propose a motion compensation algorithm specifically for near-range SAR applications. In order to achieve this goal, the aforementioned lacks of the study are covered in this thesis. The outline of the thesis is organized as follows.

In Chapter 2, restrictions on radar parameters and the primary performances of SAR system for near-range applications are given. Two frequency-domain SAR algorithms, RDA and RMA, are presented while the deduction of RDA serves as an analytical basis for the investigation of azimuth motion errors presented in Chapter 3. Their abilities to accommodate the antenna beamwidth are compared. In addition, a new metric to select the optimum antenna beamwidth is proposed, which shows better utilization of the transmitting and Doppler bandwidth than the traditional one.

Chapter 3, provides a comprehensive investigation of the influences of motion errors. Based on a complete dynamics model of the vehicle, both translation and rotation errors are investigated either by analytical deduction or by computer simulation. In addition, the concept of artificial motion errors, which are specific for near range applications and overlooked in the literature, is introduced.

On this ground, provided that motion errors were correctly obtained, a novel SAR motion compensation algorithm called the "octave division algorithm" is proposed in Chapter 4, which deals with the artificial motion errors specifically existing in near-range applications. Besides, in order to measure the slant range errors two novel motion measuring algorithms are proposed, which uses only low-cost MEMS IMU devices by exploiting the features of near-range applications.

Chapter 5 describes a SAR demonstrator system developed for the research. Verifications of the requirements for each component are provided.

Chapter 6 presents results of three types of SAR measurements carried out for different applications. With the proposed motion measuring and compensation algorithms, great improvement can be achieved for the Airquad-borne SAR application, which is realized for the first time.

Chapter 7 draws the conclusions of this work.

2 Synthetic Aperture Radar for Near-range Applications

Since SAR's debut in 1950s, comprehensive investigations have been made with respect to remote sensing applications. However, the feasibility of SAR for near-range applications has not been discussed until recently because of the restriction of the hardware [61, 24, 25, 35, 32]. Furthermore, since most of the design considerations and the choice of SAR algorithm in the literature remain the same as the remote sensing ones, the performance potential of near-range SAR systems is restricted.

Figure 2.1 depicts the geometries of SAR for remote sensing applications and near-range applications respectively, where Φ is the HPBW of the antenna in elevation and Ψ is the HPBW in azimuth. In remote sensing applications, the beamwidth in elevation governs the width of the imaged range swath as depicted in Figure 2.1a. As a contrast, the transmitted power determines the range swath in near-range applications as depicted in Figure 2.1b. In addition, in remote sensing applications, the antenna's azimuth beamwidth is restricted as a result of the conflict between the azimuth resolution and range swath. On the contrary, the range swath in near-range application is generally much shorter, e.g. less than 20 m. Therefore a higher azimuth resolution can be achieved by using a wide beamwidth antenna, e.g. the antenna beamwidth can be larger than $100°$ when an automotive anti-collision radar is used for SAR imaging. Consequently, the prerequisites of remote sensing SAR algorithms may not be met anymore. Besides, in most cases the near-range SAR system is expected to generate the image in real-time. Therefore a comprehensive consideration of SAR algorithms is needed taking into account the computational burden and applicability.

In this chapter, a SAR system with antennas of arbitrary beamwidth for near-range applications is considered. The approximations from the remote sensing applications are not applied here anymore. The geometry of near-range SAR application are first described in section 2.1. Then, the models of the transmitted, received and intermediate frequency signal for a FMCW radar scheme are introduced in section 2.2. Based on

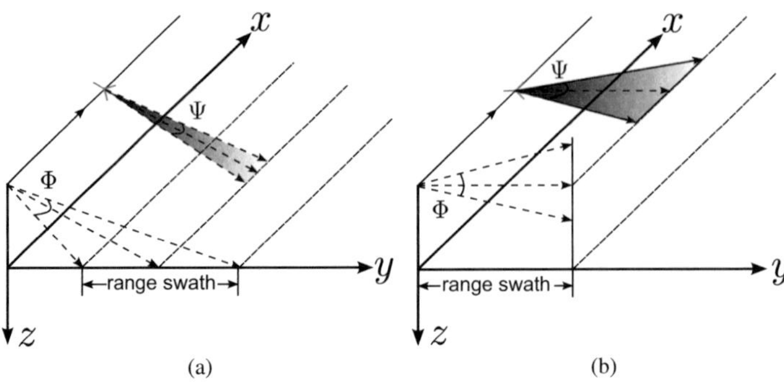

Figure 2.1: Comparison of the geometry of SAR applications: (a) remote sensing; (b) near-range.

those, the restrictions on radar parameters and the primary performances of the SAR system are developed in section 2.3. Two frequency-domain SAR algorithms, RDA and RMA, are presented respectively in subsection 2.4.1 and 2.4.2, both of them take into account the non-stop-and-go signal model specifically for FMCW radar scheme. Their abilities to accommodate the antenna beamwidth are compared in subsection 2.4.3, where a new metric to select the optimum antenna beamwidth is proposed.

2.1 Near-range Data Collection Geometry

Similar to the remote sensing SAR, the 3D data collection geometry for near-range applications can be mapped into a 2D data collection plane assuming no motion errors, with x axis representing azimuth direction, and y axis representing the range direction. Figure 2.2 depicts the geometry of a SAR working in stripmap mode.

In the ideal case, the radar traverses a straight-line path along the x axis at constant velocity, v, the instantaneous coordinates of the antenna are $(x_a, 0)$. In the following discussion, the antenna is assumed to be directed perpendicularly to the trajectory, i.e. the squint angle is set to be zero for simplicity. The shadow area is the imaged scene with length of x_{SC} and width of y_{SC}. Then the synthetic aperture interval over which

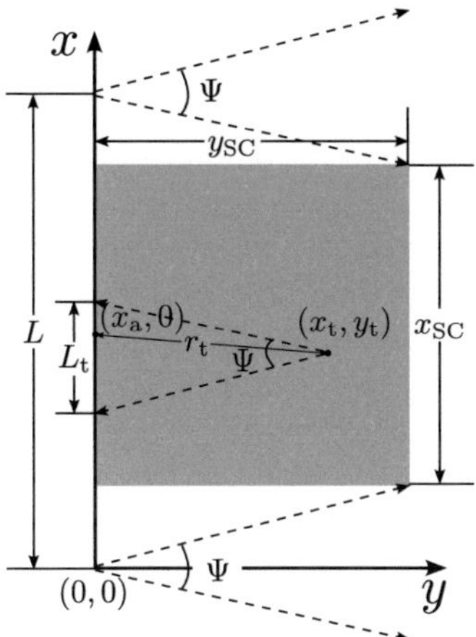

Figure 2.2: Geometry of a near-range SAR.

the stripmap SAR data contains contributions from the imaged area is

$$x \in [0, L],\qquad(2.1)$$

with

$$L = x_{\mathrm{SC}} + 2 \cdot y_{\mathrm{SC}} \cdot \tan\left(\frac{\Psi}{2}\right).\qquad(2.2)$$

A point target's aperture length, L_t, is the interval within which the target is illuminated by the antenna's azimuth half power beamwidth, Ψ, which can be calculated as

$$L_t = 2 \cdot y_t \cdot \tan\left(\frac{\Psi}{2}\right).\qquad(2.3)$$

The slant range, r_t, between the radar and a target located at (x_t, y_t) varies as a function of the antenna's along-track position x_a as

$$r_t = \sqrt{(x_a - x_t)^2 + y_t^2}.$$

When this function is at its minimum, it is called the range of the closest approach, represented by $r_{t,0}$, in the 2D representation, $r_{t,0} = y_t$.

2.2 Signal Model

The Frequency Modulated Continuous Waveform (FMCW) radar scheme is chosen for its ease of signal generation and processing. The transmitted signal can be expressed in a complex exponential form as

$$s_t(m,t) = A_0 w_r\left(t - mT_p\right) \exp\left\{ j\left[2\pi f_{min}t + \pi\gamma\left(t - mT_p\right)^2\right]\right\}. \tag{2.4}$$

Here, A_0 is the constant amplitude of the transmitted function; $w_r(t)$ denotes the envelope of the chirp, which is usually a rectangle; the quantity m is the index of the chirp with $m = 0$ representing the first transmitted chirp; the expression $t - mT_p$ represents fast-time, that is, time within a single chirp, where T_p is the chirp length; f_{min} is the minimum frequency of the chirp; $\gamma = \frac{B}{T_p}$ is the chirp rate, where B is the bandwidth of the chirp.

Defining $\hat{t}$ as the remainder of t divided by T_p, i.e.

$$\hat{t} = t \bmod T_p, \tag{2.5}$$

the transmitted signal becomes

$$s_t(m,t) = A_0 w_r(\hat{t}) \cdot \exp\left[j\left(2\pi f_{min}t + \pi\gamma\hat{t}^2\right)\right]. \tag{2.6}$$

Then the m_{th} chirp reflected from a point target located at (x_t, y_t) can be represented as

$$s_r(m,t) = A_t \cdot w_r(\hat{t} - t_d) \cdot w_a(x_a - x_t, y_t) \cdot \exp\left\{ j\left[2\pi f_{min}(t - t_d) + \pi\gamma(\hat{t} - t_d)^2\right]\right\}. \tag{2.7}$$

Here, A_t includes $\sqrt{\sigma_t}$, where σ_t is the target's radar cross-section, w_a denotes the azimuth envelope of the reflected signal, which is dictated by antenna's beam pattern and referenced to the target's coordinates, and $t_d = \frac{2r_t}{c}$ is the time delay between the transmitted and received chirp, with

$$\begin{aligned} r_t &= \sqrt{(x_a - x_t)^2 + y_t^2} \\ &= \sqrt{(x_{am} + v\hat{t} - x_t)^2 + y_t^2} \\ &= \sqrt{(m \cdot v \cdot T_p + v\hat{t} - x_t)^2 + y_t^2}. \end{aligned} \tag{2.8}$$

In the FMCW SAR case, the term $v\hat{t}$ indicates that r_t changes within a chirp duration, which is a main difference from the pulse radar where the stop-and-go approximation is applied. As a result, the SAR algorithms used in pulse radar scheme need to be modified to accommodate this extra term $v\hat{t}$, which is discussed in section 2.4.

To ease the requirements on analog-to-digital converter (ADC) and the following digital hardware, dechirp-on-receive is used to reduce the signal bandwidth before ADC. Dechirp-on-receive means to multiply the received chirps by a reference chirp, which is a replica of the transmitted signal with inverted modulation rate

$$s_{\text{ref}}(m,t) = A_0 w_r(\hat{t}) \cdot \exp\left[-j\left(2\pi f_{\min}t + \pi\gamma\hat{t}^2\right)\right]. \tag{2.9}$$

Assuming the amplitude of the transmitted signal is a rectangular shape function, the dechirped IF signal can be obtained by mixing (2.7) with (2.9) as

$$\begin{aligned} s_{\text{IF}} &= s_r \cdot s_{\text{ref}} \\ &= A_t w_r(\hat{t} - t_d) w_a(x_a - x_t, y_t) \exp\left[-j\left(2\pi f_{\min}t_d + 2\pi\gamma t_d\hat{t} - \pi\gamma t_d^2\right)\right]. \end{aligned} \tag{2.10}$$

For reasons of clarity, all the constant amplitude components are absorbed into A_t. The term $\pi\gamma t_d^2$ is an unwanted phase effect known as residual video phase (RVP), which is caused by performing dechirp in the time domain and should be removed using the operation called range deskew [18]. Since in near-range SAR applications RVP is far smaller than the other two terms in the phase, i.e.

$$\frac{2\pi f_{\min}t_d + 2\pi\gamma t_d\hat{t}}{\pi\gamma t_d^2} = \frac{\frac{2f_{\min}}{B} \cdot T_p + 2\hat{t}}{t_d} \gg 1, \tag{2.11}$$

it is reasonable to leave RVP out in the subsequent equations.

Combining (2.10) and (2.8), the intermediate frequency (IF) signal can be expressed as a function of x_{am} and $\hat{t}$

$$\begin{aligned} s_{\text{IF}}(x_{\text{am}}, \hat{t}) &= A_t w_r(\hat{t} - t_d) w_a(x_a - x_t, y_t) \\ &\quad \cdot \exp\left[-j\frac{4\pi}{c}(f_{\min} + \gamma\hat{t})\sqrt{(x_{\text{am}} + v\hat{t} - x_t)^2 + y_t^2}\right]. \end{aligned} \tag{2.12}$$

Equation (2.12) is the base of all the following analysis, different SAR algorithms presented in this thesis are developed from this model and the influence of motion error is also studied via this model.

2.3 Parameters and Performances

In this section, the basic parameters of the SAR are discussed and designed specifically for near-range SAR applications, and the range and azimuth resolutions are deducted accordingly.

2.3.1 Squint Angle

In this thesis, the squint angle of the SAR is assumed to be zero. When the squint angle becomes large, the SAR algorithm needs to be adapted as discussed in [19]. However, the motion compensation algorithms presented remain unchanged.

2.3.2 Azimuth HPBW

The antenna's azimuth HPBW, Ψ, is inversely proportional to the azimuth resolution, Δr_{azi}, of SAR. However, the improvement of Δr_{azi} is in conflict with the increased swath width in remote SAR applications via the choice of pulse repetition frequency (PRF) [109]. For near-range SAR applications this constraint on Ψ is greatly relaxed because of the short swath width, which is demonstrated with a case in the following discussion of transmitted chirp length. Therefore, in the following derivations of radar's parameters and performances no approximations based on a small Ψ as in remote sensing SAR are applied.

2.3.3 Polarization

Though the four modes of radar transmit-receive polarization, which are referred to as HH, HV, VH, and VV, provide a vector information base to characterize the target signature in a SAR system, the analysis in this thesis uses scalar wave theory to simplify

14

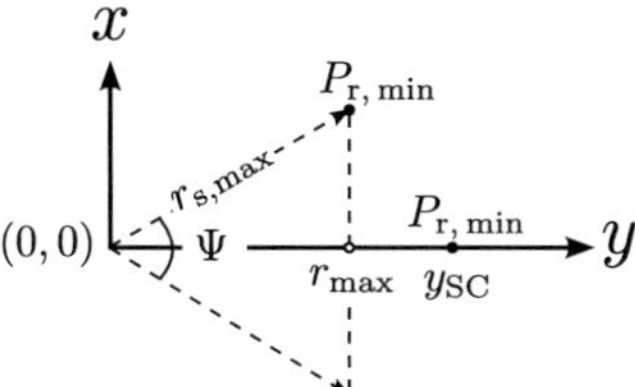

Figure 2.3: Geometry of $r_{\max}$ and y_{SC}: at the locations represented by solid dot, the received power of the same target equals to the minimum detectable power of the receiver.

the notation. However, all the results and discussions on the scalar wave theory can be easily adapted for the vector wave theory.

2.3.4 Maximum Working Range of SAR

The maximum working range of SAR, $r_{\max}$, is defined as the range of the closest approach to the furthest target of interest. On the one hand, it determines the transmitted power of the radar, P_t, when designing a specific near-range SAR. On the other hand, it can be determined by the working range of a universal radar.

As depicted in Figure 2.3, assuming the antenna is located at point $(0,0)$ and considering only the area within the HPBW of antenna, the maximum slant range, $r_{s,\max}$, of the furthest target of interest in SAR application is a function of $r_{\max}$ and Ψ as

$$r_{s,\max} = \frac{r_{\max}}{\cos\left(\frac{\Psi}{2}\right)}. \tag{2.13}$$

According to the radar equation, the minimum detectable received power, $P_{r,\min}$, reflected from the furthest target of interest can be calculated when the target is just at the edge of the HPBW as

$$P_{r,\min} = P_r\left(\frac{\Psi}{2}, r_{s,\max}\right) = P_t \cdot \frac{G_t\left(\frac{\Psi}{2}\right) G_r\left(\frac{\Psi}{2}\right)\lambda^2}{(4\pi)^3 \cdot r_{s,\max}^4} \cdot \sigma_t, \tag{2.14}$$

where P_t is the transmitted power, G_t and G_r are the gain of the transmitting and receiving antennas respectively, which are the functions of the angle of incidence. Given the requirement of r_{max} and Ψ and the performance of the receiver, P_t can be calculated according to (2.14).

As the target moves away from the antenna along the y axis, the returning power, $P_r(0, y_t)$, becomes weaker until less than $P_{r,min}$, from where the target is invisible to the radar. Therefore the maximum working range of the radar can be defined via

$$P_r(0, y_{SC}) = P_r\left(\frac{\Psi}{2}, r_{s,max}\right).$$

(2.15)

Given $G_t\left(\frac{\Psi}{2}\right) = \frac{1}{2}G_t(0)$ and $G_r\left(\frac{\Psi}{2}\right) = \frac{1}{2}G_r(0)$, y_{SC} can be calculated as

$$y_{SC} = \frac{\sqrt{2}}{\cos\left(\frac{\Psi}{2}\right)} \cdot r_{max}.$$

(2.16)

With (2.16), r_{max} of a SAR system can be estimated when the maximum working range of the radar is known, e.g. in the case that an already existing automotive radar is used for additional SAR applications.

2.3.5 Transmitted Chirp Length

The transmitted chirp length T_p is essentially the reciprocal of the space sampling frequency, i.e. PRF, in pulse radar scheme. To meet the Nyquist criterion, the PRF should be larger than the Doppler bandwidth B_{Dop} (as the SAR corresponds to a complex sampler in azimuth direction) multiplied by an azimuth oversampling factor, $\alpha_{os,azi}$, which is usually between 1.1 and 1.4. The Doppler frequency f_{Dop} caused by the relative movement between the radar and the target (located at range bin of y_t) can be calculated as

$$
\begin{aligned}
f_{Dop}(t) &= \frac{1}{2\pi} \cdot \frac{d}{dt}\left[2\pi f \cdot \frac{2\sqrt{y_t^2 + (vt)^2}}{c}\right] \\
&= \frac{2v \cdot vt}{\lambda \cdot \sqrt{y_t^2 + (vt)^2}} \\
&= \frac{2v}{\lambda} \cdot \sin(\psi),
\end{aligned}
$$

(2.17)

(2.18)

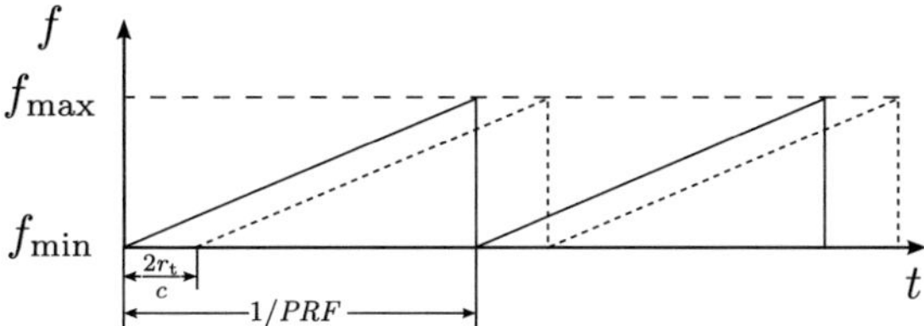

Figure 2.4: The transmitted chirp and its reflected chirps from a certain target.

where λ is the wavelength of the instantaneous transmitted frequency, v is the azimuth velocity of the radar, and ψ is the angle between the instantaneous slant range vector and the zero Doppler plane. To simplify the calculation, the target is assumed to be only illuminated by the antenna within the HPBW. The period of when a certain target is illuminated by the antenna is called synthetic aperture time, T_{L_t}, which can be calculated as

$$T_{L_t} = \frac{L_t}{v} = \frac{2 \cdot y_t \cdot \tan\left(\frac{\Psi}{2}\right)}{v}.$$ (2.19)

Then B_{Dop} can be calculated accordingly

$$\begin{aligned} B_{\mathrm{Dop}} &= f_{\mathrm{Dop}}\left(\frac{T_{L_t}}{2}\right) - f_{\mathrm{Dop}}\left(-\frac{T_{L_t}}{2}\right) \\ &= \frac{4v\sin\left(\frac{\Psi}{2}\right)}{\lambda}. \end{aligned}$$ (2.20)

So the upper limit of T_{p} is

$$T_{\mathrm{p,max}} = \frac{1}{\alpha_{\mathrm{os,azi}} \cdot B_{\mathrm{Dop}}} = \frac{\lambda}{\alpha_{\mathrm{os,azi}} \cdot 4v\sin\left(\frac{\Psi}{2}\right)}.$$ (2.21)

On the other hand, in remote sensing SAR applications, the lower limit of T_{p} is restricted by the slant range of $\frac{1}{PRF} \cdot \frac{c}{2}$, which is decided both by the height of the radar and Θ . The *PRF* should be low enough so that most or all of the near range to far range interval illuminated by the beam (the swath width) fall within the receive window. Similarly, in near-range SAR applications, T_{p} should be long enough so that most of the energy reflected by the furthest target falls within the same transmitted chirp, as illustrated in Figure 2.4.

Since the antenna is directed approximately in parallel with the ground, the swath width in these cases is decided by P_t, which should be designed according to the furthest slant range $r_{s,max}$ of the specific application. Therefore, the lower limit of T_p can be calculated as

$$T_{p,min} = \frac{1}{1-80\%} \cdot \frac{2r_{s,max}}{c},$$ (2.22)

where 80% of the energy reflected from the furthest target fall within the same transmitted chirp.

$T_{p,max}$ is preferred since the data rate of SAR is inversely proportional to T_p. Generally, $T_{p,max}$ is much higher than $T_{p,min}$, e.g. $\frac{T_{p,max}}{T_{p,min}} > 3000$, when $\lambda_c = 1.25\,\mathrm{cm}$ (the wavelength of the center frequency of chirp), $v = 4\,\mathrm{m/s}$, $\Psi = 30°$ and $r_{max} = 20\,\mathrm{m}$. Therefore once $T_{p,max}$ is determined, the targets with $y_t > r_{max}$ which are illuminated un-intentionally (though the reflected power from them is limited as a result of the designed transmitted power) will not be aliased into the next chirp.

2.3.6 Sampling Frequency of IF Signal

Since the dechirp-on-receive scheme is employed, the dechirped signal is sampled by a ADC, and the sampling frequency, f_s, should meet the Nyquist criterion. For real sampling f_s should be higher than two times of the maximum frequency of s_{IF}, which can be calculated as

$$\begin{aligned}
f_{IF,max} &= \max\left\{ \frac{d}{dt}\left[\frac{2}{c}(f_{min} + \gamma\hat{t})\sqrt{(x_{am} + v\hat{t} - x_t)^2 + y_t^2} \right] \right\} \\
&= \frac{2}{c} \cdot \max\left[\gamma \cdot r_t \cdot \frac{d\hat{t}}{dt} + (f_{min} + \gamma\hat{t}) \cdot \frac{dr_t}{dt} \right] \\
&= \frac{2}{c} \cdot \left[\gamma \cdot y_{SC} + f_{max} \cdot v \cdot \sin\left(\frac{\Psi}{2}\right) \right].
\end{aligned}$$ (2.23)

The first term in the square brackets denotes the position information and the second term denotes the velocity information. It is clear that f_s is determined by y_{SC}, Ψ and v. Generally, the first term in the square brackets is much larger than the second term because of the low vehicle's velocity in the near-range applications. Therefore, f_s is determined by y_{SC} primarily.

2.3.7 Range Resolution

The 3 dB range resolution, Δr_{ran}, of the FMCW radar is governed by B as

$$\Delta r_{\text{ran}} = k_{\text{win}}^{\text{ran}} \cdot \frac{c}{2B}, \tag{2.24}$$

where $k_{\text{win}}^{\text{ran}}$ is a broadening factor related to the weighting window applied on the range frequency direction, e.g. for a rectangular windows, $k_{\text{win}}^{\text{ran}} = 0.89$, and for a Taylor window with sidelobe level (SLL) of -35 dB, and 5 nearly constant-level sidelobes adjacent to the mainlobe, $k_{\text{win}}^{\text{ran}} = 1.19$.

2.3.8 Azimuth Resolution

The 3 dB azimuth resolution, Δr_{azi}, of the SAR is independent of the relative range between the radar and the target. It is inversely proportional to the Doppler bandwidth, and can be calculated as

$$\Delta r_{\text{azi}} = k_{\text{ant}} \cdot k_{\text{win}}^{\text{azi}} \cdot \frac{v}{B_{\text{Dop}}} = k_{\text{ant}} \cdot k_{\text{win}}^{\text{azi}} \cdot \frac{\lambda_c}{4 \sin\left(\frac{\Psi}{2}\right)}, \tag{2.25}$$

where $k_{\text{win}}^{\text{azi}}$ is a broadening factor related to the windows applied on the azimuth frequency direction, and k_{ant} is a factor related to the azimuth beam pattern.

2.4 Comparison of SAR Algorithms

Restricted by the realistic hardware environment, the most accurate time-domain based SAR algorithms cannot be applied in real-time applications yet. Thus the choice of SAR algorithms is limited in frequency-domain based algorithms, such as CSA, RDA and RMA. In pulse radar scheme CSA has better performance than RDA even with less computational burden [98]. On the contrary, as indicated in [99], it requires extra processing which balances out its pros when applied in the radar with dechirp-on-receive scheme. Therefore RDA is chosen here as a representation of SAR algorithms with a light computational burden, albeit with less accuracy, to compare with the most computationally expensive and accurate SAR algorithm, RMA. Though a

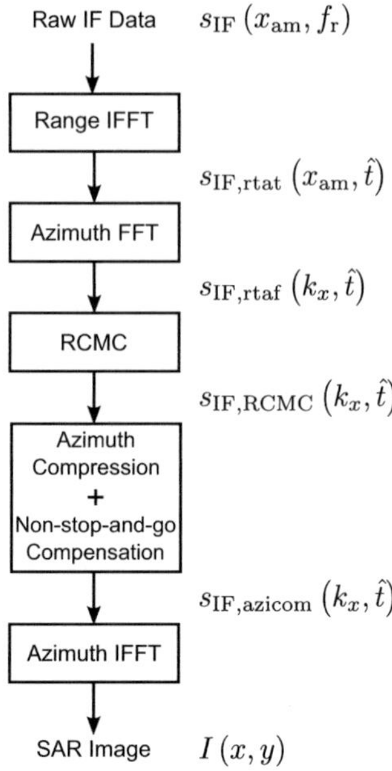

Figure 2.5: Block diagram of RDA.

detailed comparison between these two algorithms has been presented in [110], two exclusive considerations are included here: the influence of continuous movement of antenna within one chirp, and the algorithm's adaptability for the usage of wide beamwidth antenna. Since both algorithms are derived from the same recorded raw IF data form (2.12), it is convenient for later chapters to analyze the influences of motion errors on the Stolt mapping stage in RMA by referring to its corresponding stages in RDA.

2.4.1 Range Doppler Algorithm (RDA)

RDA processes the raw radar data with two steps for range and azimuth separately by approximating the hyperbolic range equation into a parabolic equation. The block diagram of RDA is shown in Figure 2.5.

The dechirp-on-receive approach described in section 2.2 not only accomplishes bandwidth reduction but also performs the range compression operation in the pulse scheme SAR. The raw IF data, $s_{IF}(x_{am}, \hat{t})$, is actually in the range-frequency azimuth-time domain. In order to perform range cell migration correction (RCMC), $s_{IF}(x_{am}, \hat{t})$ needs to be transformed to range-time domain by performing range inverse fast Fourier transform (IFFT). Using f_r to denote the demodulated range frequency as

$$f_r = f_{min} + \gamma\hat{t} - f_c, \tag{2.26}$$

where f_c is the center frequency of the chirp, $s_{IF}(x_{am}, \hat{t})$ can be written as

$$s_{IF}(x_{am}, f_r) = A_t \cdot w_a(x_a - x_t, y_t) \cdot w_r(f_r) \cdot \exp\left(-j2\pi f_r \frac{2r_t}{c} - j\frac{4\pi}{c}f_c r_t\right), \tag{2.27}$$

where the delay t_d is omitted from the range frequency envelope, $w_r(\cdot)$, since it is much smaller than T_p as discussed in subsection 2.3.5. Furthermore, a range frequency window can be applied on $s_{IF}(x_{am}, f_r)$ at this step if needed. For reasons of clarity, the window is absorbed into $w_r(f_r)$.

The range IFFT of $s_{IF}(x_{am}, f_r)$ can be performed as

$$\begin{aligned}
s_{IF,rtat}(x_{am}, \hat{t}) &= A_t w_a(x_a - x_t, y_t) \exp\left(-j\frac{4\pi}{c}f_c r_t\right) \\
&\quad \cdot \int_{-\frac{B}{2}}^{\frac{B}{2}} w_r(f_r) \cdot \exp\left(-j2\pi f_r \frac{2r_t}{c} + j2\pi f_r \hat{t}\right) df_r \\
&\approx A_1 \cdot p_r\left(\hat{t} - \frac{2r_t}{c}\right) \cdot w_a(x_a - x_t, y_t) \cdot \exp\left(-j\frac{4\pi}{c}f_c r_t\right),
\end{aligned} \tag{2.28}$$

where the subscript "rtat" denotes "range-time azimuth-time, A_1 denotes the over all gain which includes σ_t, and the range envelope $p_r\left(\hat{t} - \frac{2r_t}{c}\right)$ is the inverse Fourier transform of $w_r(f_r)$, which incorporates the target range migration via the azimuth varying parameter $\frac{2r_t}{c}$.

The continuous movement of the antenna during one chirp period causes the term $v\hat{t}$ appearing in r_t and makes it also be a function of f_r, and therefore "$\approx$" is used in (2.28). However, according to (2.21), $v\hat{t}$ is limited to a very small value as

$$v\hat{t} < vT_{\mathrm{P}} < \frac{\lambda}{\alpha_{\mathrm{os,azi}} \cdot 4\sin\left(\frac{\Psi}{2}\right)}.$$ (2.29)

For a SAR system operating at 24 GHz with $\Psi = 30°$, $v\hat{t}$ is less than 1 cm, which is significantly smaller than the r_{t} of interest. Therefore its effect on the range IFFT can always be ignored.

The azimuth fast Fourier transform (FFT) with respect to x_{am} is given by

$$s_{\mathrm{IF,rtaf}}\left(k_x,\hat{t}\right) = \int s_{\mathrm{IF,rtat}}\left(x_{\mathrm{am}},\hat{t}\right) \cdot \exp\left(-jk_x x_{\mathrm{am}}\right) dx_{\mathrm{am}},$$ (2.30)

where the subscript "rtaf" denotes "range-time azimuth-frequency", and the azimuth spatial frequency k_x is defined as

$$k_x = \frac{2\pi f_{\mathrm{Dop}}}{v}.$$ (2.31)

The closed form solution can be derived using the method of principle of stationary phase (POSP). The phase of the Fourier integrand is

$$\phi\left(x_{\mathrm{am}}\right) = -\frac{4\pi}{\lambda_{\mathrm{c}}}r_{\mathrm{t}} - k_x x_{\mathrm{am}},$$ (2.32)

where λ_{c} is the wavelength of the center frequency. By approximating the range equation from (2.8) to a parabolic equation as

$$r_{\mathrm{t}} \approx y_{\mathrm{t}} + \frac{\left(x_{\mathrm{am}} + v\hat{t} - x_{\mathrm{t}}\right)^2}{2y_{\mathrm{t}}},$$ (2.33)

the derivative of $\phi\left(x_{\mathrm{am}}\right)$ with respect to x_{am} can be calculated as

$$\frac{d\phi_x\left(x_{\mathrm{am}}\right)}{x_{\mathrm{am}}} = -\pi \cdot \frac{4}{\lambda_{\mathrm{c}}y_{\mathrm{t}}} \cdot \left(x_{\mathrm{am}} + v\hat{t} - x_{\mathrm{t}}\right) - k_x,$$ (2.34)

which is zero when

$$x_{\mathrm{am}}^* = -\frac{\lambda_{\mathrm{c}}y_{\mathrm{t}}}{4\pi} \cdot k_x + x_{\mathrm{t}} - v\hat{t}.$$ (2.35)

By substituting (2.35) into (2.30), the IF data after azimuth FFT can be obtained

$$s_{\text{IF,rtaf}}(k_x,\hat{t}) = A_2 \cdot p_r\left(\hat{t} - \frac{2r_t(k_x)}{c}\right) \cdot W_a(k_x) \cdot \exp\left(-j\frac{4\pi f_c}{c} y_t\right)$$

$$\cdot \exp\left(\frac{\lambda_c y_t}{8\pi} \cdot k_x^2 - k_x \cdot x_t + k_x \cdot v\hat{t}\right), \tag{2.36}$$

where A_2 denotes the overall gain which includes σ_t. The azimuth envelope, $w_a(x_a - x_t, y_t)$, is transformed into $W_a(k_x)$ with its shape preserved as

$$W_a(k_x) = w_a\left(-\frac{\lambda_c y_t}{4\pi} \cdot k_x, y_t\right). \tag{2.37}$$

The azimuth frequency window is applied at this step and for reasons of clarity absorbed into W_a.

The range cell migration term, RCM, in $p_r\left(\hat{t} - \frac{2r_t(k_x)}{c}\right)$ can be obtained by substituting (2.35) into (2.33) as

$$r_t(k_x) = y_t + \frac{\lambda_c^2 y_t}{32\pi^2} k_x^2 = y_t + RCM. \tag{2.38}$$

After performing RCMC using the method based on sinc interpolation, the signal becomes

$$s_{\text{IF,RCMC}}(k_x,\hat{t}) = A_2 \cdot p_r\left(\hat{t} - \frac{2y_t}{c}\right) \cdot W_a(k_x) \cdot \exp\left(-j\frac{4\pi f_c}{c} y_t\right) \cdot \exp\left(-j \cdot k_x x_t\right)$$

$$\cdot \exp\left[j\left(\frac{\lambda_c y_t}{8\pi} \cdot k_x^2 + k_x \cdot v\hat{t}\right)\right]. \tag{2.39}$$

The matched filter for azimuth compression is the conjugate of the last exponential term in (2.39) as

$$H_{\text{azi}}(k_x,\hat{t}) = \exp\left[-j\left(\frac{\lambda_c y_t}{8\pi} \cdot k_x^2 + k_x \cdot v\hat{t}\right)\right], \tag{2.40}$$

which also compensates the effect of the continuous moving of antenna within one chirp by the term $k_x \cdot v\hat{t}$. After azimuth compression the signal can be expressed as

$$
\begin{aligned}
s_{\text{IF,azicom}}\left(k_x,\hat{t}\right) &= s_{\text{IF,RCMC}}\left(k_x,\hat{t}\right) \cdot H_{\text{azi}}\left(k_x,\hat{t}\right) \\
&= A_2 \cdot p_{\text{r}}\left(\hat{t}-\frac{2y_{\text{t}}}{c}\right) \cdot W_{\text{a}}\left(k_x\right) \cdot \exp\left(-j \cdot k_x x_{\text{t}}\right) \\
&\quad \cdot \exp\left(-j\frac{4\pi f_{\text{c}}}{c}y_{\text{t}}\right).
\end{aligned}
\tag{2.41}
$$

By performing azimuth IFFT on (2.41) , the compressed SAR image can be obtained

$$
\begin{aligned}
I\left(x,y\right) &= \int s_{\text{IF,rtaf,azicom}}\left(k_x,\hat{t}\right) \cdot \exp\left(jk_x x_{\text{am}}\right) dx_{\text{am}} \\
&= A_2 \cdot p_{\text{r}}\left(y-y_{\text{t}}\right) \cdot \int W_{\text{a}}\left(k_x\right) \cdot \exp\left(-j \cdot k_x x_{\text{t}}\right) \cdot \exp\left(jk_x x_{\text{am}}\right) dx_{\text{am}} \\
&\quad \cdot \exp\left(-j\frac{4\pi f_{\text{c}}}{c}y_{\text{t}}\right) \\
&= A_3 \cdot p_{\text{r}}\left(y-y_{\text{t}}\right) \cdot p_{\text{a}}\left(x-x_{\text{t}}\right) \cdot \exp\left(-j\frac{4\pi f_{\text{c}}}{c}y_{\text{t}}\right),
\end{aligned}
\tag{2.42}
$$

where A_3 contains the overall gain including σ_{t}. The range envelope $p_{\text{r}}\left(\cdot\right)$ is now expressed as a function y instead of $\hat{t}$, and subscript "$_{\text{am}}$" in x_{am} is omitted for reasons of clarity. The azimuth envelope $p_{\text{a}}\left(x\right)$ is the inverse Fourier transform of $W_{\text{a}}\left(k_x\right)$. The exponential term, $\exp\left(-j\frac{4\pi f_{\text{c}}}{c}y_{\text{t}}\right)$, being a function of y_{t} is important in interferometry and polarimetry applications but not important in intensity images.

2.4.2 Range Migration Algorithm (RMA)

The RMA is the most accurate frequency-domain SAR algorithm, which can correct the range varying range cell migration over wide aperture due to the use of the accurate hyperbolic form of range equation. On the contrary, the calculus of interpolation in the Stolt mapping makes RMA the most computationally expensive SAR algorithm in the frequency-domain. Since the assumption that the effective radar velocity is constant along range direction holds for the near-range applications, traditional RMA is

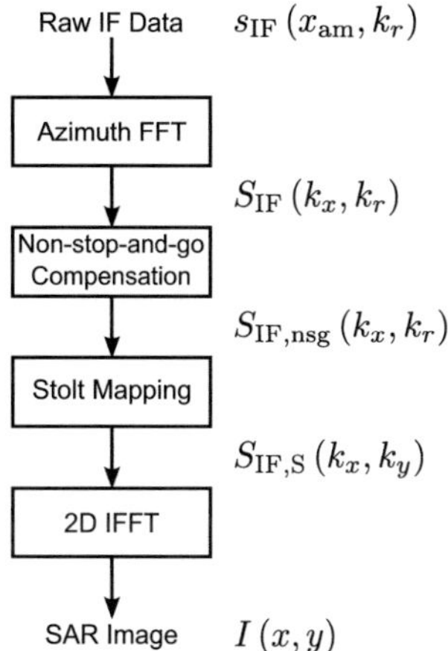

Figure 2.6: Block diagram of RMA.

adopted here instead of the modified versions, which take into account the change of the effective radar velocity as proposed in [111, 112].

The block diagram of RMA is shown in Figure 2.6.

As discussed in 2.4.1, the raw IF data is actually in the range-frequency azimuth-time domain. By substituting the spatial frequency of slant range, $k_r = \frac{4\pi}{c}(f_{\min} + \gamma\hat{t})$, into (2.12), the raw data is expressed in the range-spatial-frequency azimuth-time domain as

$$s_{\mathrm{IF}}(x_{\mathrm{am}}, k_r) = A_t w_r(k_r) w_a(x_a - x_t, y_t) \exp\left[-jk_r \cdot \sqrt{(x_{\mathrm{am}} + v\hat{t} - x_t)^2 + y_t^2}\right], \quad (2.43)$$

where the delay t_{d} is omitted from the range amplitude function $w_r(\cdot)$ since it is much smaller than the chirp length and $w_r(\cdot)$ is now expressed as a function of k_r. Like RDA, the range frequency window, if applied, is absorbed into $w_r(\cdot)$ for reasons of clarity.

Performing a range Fourier transform on (2.43) with respect to the variable x_{am}, the IF signal is then transformed into the 2D spatial frequency domain as

$$S_{\mathrm{IF}}(k_x, k_r) = A_t w_r(k_r) \cdot \int w_a(x_a - x_t, y_t) \exp[\phi(x_{\mathrm{am}})] dx_{\mathrm{am}}. \quad (2.44)$$

Again the closed solution of (2.44) can be obtained by using the method of POSP. The phase of the Fourier integrand is

$$\phi\left(x_{am}\right) = -k_r \cdot \sqrt{\left(x_{am} + v\hat{t} - x_t\right)^2 + y_t^2} - k_x x_{am}.\tag{2.45}$$

The derivative of $\phi\left(x_{am}\right)$ with respect to x_{am} can be obtained by

$$\frac{d\phi\left(x_{am}\right)}{dx_{am}} = -\frac{k_r\left(x_{am} + v\hat{t} - x_t\right)}{\sqrt{\left(x_{am} + v\hat{t} - x_t\right)^2 + y_t^2}} - k_x,\tag{2.46}$$

which is zero when

$$x_{am}^* = -\frac{y_t k_x}{\sqrt{k_r^2 - k_x^2}} + x_t - v\hat{t}.\tag{2.47}$$

By substituting (2.47) into (2.44), the 2D spatial spectrum can be expressed as

$$\begin{aligned}
S_{IF}\left(k_x, k_r\right) &= A_1 w_r\left(k_r\right) W_a\left(k_x, k_r\right) \\
&\quad \cdot \exp\left[j\left(-k_x \cdot x_t - \sqrt{k_r^2 - k_x^2} \cdot y_t + k_x \cdot v\hat{t}\right)\right],
\end{aligned}\tag{2.48}$$

where A_1 denotes the overall gain which includes σ_t. The azimuth envelope of the IF data, $w_a\left(x_a - x_t, y_t\right)$, is transformed into $W_a\left(k_x, k_r\right)$ with its shape preserved basically as

$$W_a\left(k_x, k_r\right) = w_a\left(-\frac{y_t k_x}{\sqrt{k_r^2 - k_x^2}}, y_t\right),\tag{2.49}$$

which is now also a function of k_r. Different from a rectangular support band in $\left(x_{am}, k_r\right)$ domain before azimuth Fourier transform, the 2D support band of the radar signal in $\left(k_x, k_r\right)$ domain turns into a trapezoid, where the bandwidth of k_x is proportional to k_r. Like RDA, the azimuth frequency window is applied at this step and for reasons of clarity absorbed into W_a.

As a result of the continuous antenna movement within one chirp period, the term $k_x \cdot v\hat{t}$ appears in (2.48). However, the effect of continuous motion within one chirp period can be compensated by multiplying (2.48) with $\exp\left(-jk_x \cdot v\hat{t}\right)$ as

$$\begin{aligned}
S_{IF,nsg}\left(k_x, k_r\right) &= S_{IF}\left(k_x, k_r\right) \cdot \exp\left(-jk_x \cdot v\hat{t}\right) \\
&= A_1 w_r\left(k_r\right) W_a\left(k_x, k_y\right) \exp\left[-j\left(k_x \cdot x_t + \sqrt{k_r^2 - k_x^2} \cdot y_t\right)\right].
\end{aligned}\tag{2.50}$$

By changing the variables as

$$k_y = \sqrt{k_r^2 - k_x^2},\tag{2.51}$$

$S_{IF,nsg}(k_x, k_r)$ becomes a 2D spectrum with linear phase terms in both dimensions as

$$S_{IF,S}(k_x, k_y) = A_1 w_r\left(\sqrt{k_x^2 + k_y^2}\right) W_a(k_x, k_y) \exp\left[-j(k_x x_t + k_y y_t)\right].\tag{2.52}$$

In order to obtain the impulse response function (IRF) of the target locating at (x_t, y_t) by performing 2D IFFT on (2.52), Stolt mapping is applied to resample the signal along k_r to give output samples that are uniformly spaced in k_y. Note that the slant range frequency envelope, $w_r\left(\sqrt{k_x^2 + k_y^2}\right)$, is now a function of both k_x and k_y, and the support band of the signal data is transformed from a trapezoid into an annulus. The effect of the deformation of the support band, which deteriorates the resolution of a SAR system, will be discussed in the next subsection.

At the end of the RMA, applying 2D IFFT on (2.52), the SAR image can be obtained

$$\begin{aligned}
I(x,y) &= \int S_{IF,S}(k_x, k_y) \cdot \exp\left[j(k_x x + k_y y)\right] dx dy \\
&= A_2 \cdot p_r(y - y_t) \cdot p_a(x - x_t)
\end{aligned}\tag{2.53}$$

Here, A_2 contains the overall gain including σ_t; $p_r(\cdot)$ and $p_a(\cdot)$ is the range and azimuth envelope of the IRF respectively.

2.4.3 Beamwidth Adaptability

Since the approximated parabolic range equation is used in RDA, the strong cross coupling between range and azimuth is not considered, which is a result of a wide Ψ and will defocus the final image. As shown in Figure 2.7a and 2.7c, the phase of $s_{IF,rtaf,azicom}$ becomes non-linear as Ψ increases. Consequently, more energy is leaked to the surrounding sidelobes of the IRF, as the comparison of Figure 2.7b and 2.7d indicates.

After Stolt mapping in RMA the shape of the support band in 2D spatial frequency domain is annular instead of rectangular. As shown in Figure 2.8a, the annular support band can be assumed to be rectangular only when the antenna beamwidth is small,

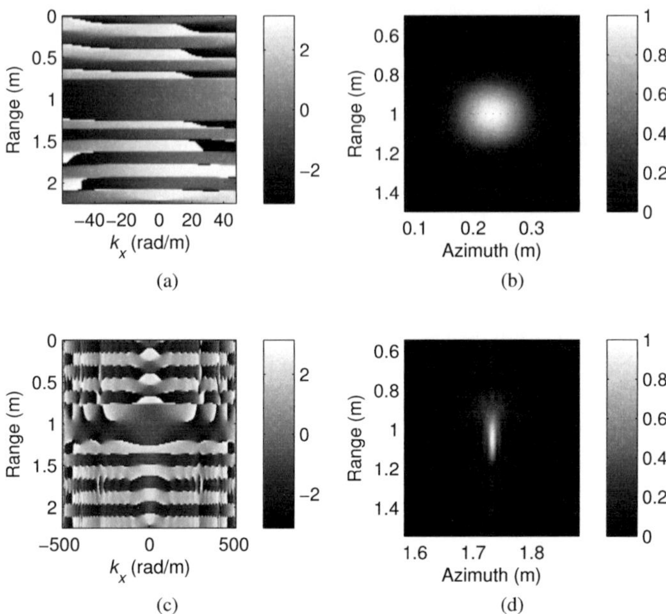

Figure 2.7: Comparison between narrow and wide Ψ processed by RDA with $f_c = 24\,\text{GHz}$ and $B = 1\,\text{GHz}$: (a) phase after azimuth compression when $\Psi = 6°$, (b) IRF when $\Psi = 6°$, (c) phase after azimuth compression when $\Psi = 60°$, (d) IRF when $\Psi = 60°$.

e.g. $\Psi < 10°$, based on which the theoretical Δr_{ran} and Δr_{azi} are derived in (2.24) and (2.25). As Ψ increases, the central angle of the annular support band become larger. Since the bandwidth of k_y increases starting from the bandwidth of k_r as Ψ increases, Δr_{ran} is better than its theoretical value calculated according to (2.24). However, since part of the assumed-to-be rectangular support band is unfilled, as shown in Figure 2.8c, along range direction only part of the azimuth mainlobe is focused using the full Doppler bandwidth. Away from the center of IRF, less Doppler bandwidth is used and consequently the broadening of azimuth mainlobe is more severely, as shown in Figure 2.8d.

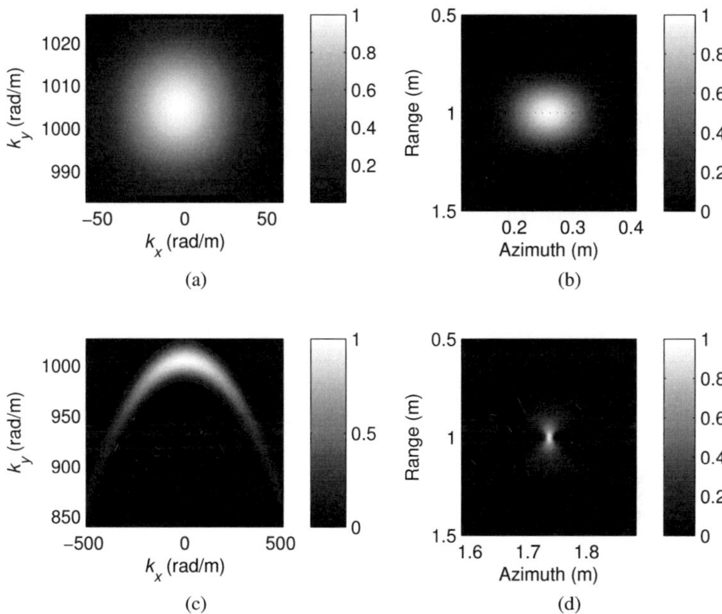

Figure 2.8: Comparison between narrow and wide Ψ processed by RMA with $f_c = 24\,\text{GHz}$ and $B = 1\,\text{GHz}$: (a) the support band in (k_x, k_y) when $\Psi = 6°$; (b) IRF when $\Psi = 6°$; (c) the support band in (k_x, k_y) when $\Psi = 60°$; (d) IRF when $\Psi = 60°$.

As depicted in Figure 2.7d and Figure 2.8d, when Ψ becomes large, the shapes of the distorted IRF look very different between RDA and RMA. In RDA the energy spreads from the mainlobe randomly to its surrounding area with Δr_{azi} remaining constant within the mainlobe. On the contrary, the contour of the mainlobe in RMA is not rectangular but presents a shape of "8". It is not easy to compare these two SAR algorithms qualitatively in terms of their adaptability to Ψ.

A new evaluation metric for the SAR algorithm's adaptability to Ψ is presented here, which is based on two parameters: integrated sidelobe ratio (*ISLR*) and integrated resolution portion ratio (*IRPR*). On the one hand, the effect of energy leaks from the mainlobe can be evaluated quantitatively by *ISLR*, which is defined as the ratio of the energy in sidelobes of the IRF to the energy in the mainlobe. In this definition the borders between mainlobe and sidelobes are the locations of the first nulls of the IRF obtained by the ideal rectangular support band associated with the aperture weighting function being used. Since the distortion of the IRF does not happen solely along azimuth or range direction, a 2D *ISLR* should be determined. On the other hand, *IRPR*, which is defined as the ratio of integrated energy in the resolution portion of the mainlobe to the whole energy of the mainlobe, is a more practical metric to define resolution when the mainlobe of the IRF is distorted.

The procedure of the metric is as follows:

1. A 2D rectangular support band is formed in (k_x, k_y) domain. The phase in the support band is set according to a certain point target's coordinates and the transmitted chirp's frequency range. First, the amplitude of the support band inherits from $S_{\mathrm{IF}}(k_x, k_r)$ in RMA, and thereby the antenna's pattern is considered. Then, a Taylor window is applied in both k_x and k_y direction. The parameters of the Taylor window are: *SLL* of $-35\,\mathrm{dB}$, which is the peak sidelobe level relative to the mainlobe peak; $\bar{n}$ of 5, which is the number of the nearly constant-level sidelobes adjacent to the mainlobe; N_{Taylor}, which is the number of coefficients determined by Ψ (signal outside Ψ is assumed to be zero in the following discussion).

2. The IRF of the ideal rectangular support band is generated using a 2D FFT, from which the following parameters can be determined and denoted with superscript "ideal": $\Delta r_{\mathrm{azi}}^{\mathrm{ideal}}$ and $\Delta r_{\mathrm{ran}}^{\mathrm{ideal}}$; $d_{\mathrm{azi}}^{\mathrm{ideal}}$ and $d_{\mathrm{ran}}^{\mathrm{ideal}}$ as the length from the center of IRF to the first null in azimuth and range directions respectively; $IRPR^{\mathrm{ideal}}$ and $ISLR^{\mathrm{ideal}}$.

3. The $\Delta r_{\mathrm{azi}}^{\mathrm{RMA}}$, $\Delta r_{\mathrm{ran}}^{\mathrm{RMA}}$ and $ISLR_{d^{\mathrm{ideal}}}^{\mathrm{RMA}}$ of RMA can be obtained, where the subscript "d^{ideal}" denotes the $ISLR^{\mathrm{RMA}}$ is calculated using divisions defined by $d_{\mathrm{azi}}^{\mathrm{ideal}}$ and $d_{\mathrm{ran}}^{\mathrm{ideal}}$. As depicted in Figure 2.8d, only part of the mainlobe is fully focused in azimuth direction. Therefore, by keeping $d_{\mathrm{ran}}^{\mathrm{ideal}}$ unchanged, d_{azi} is extended until the resultant $ISLR_{d_{\mathrm{azi}}^{\mathrm{RMA}}}^{\mathrm{RMA}}$ equals to $ISLR_{d^{\mathrm{ideal}}}^{\mathrm{ideal}}$, where the subscript $d_{\mathrm{azi}}^{\mathrm{RMA}}$ denotes the new effective division in azimuth direction. Both $d_{\mathrm{ran}}^{\mathrm{ideal}}$ and $d_{\mathrm{azi}}^{\mathrm{RMA}}$ define the effective mainlobe of the IRF of RMA. Subsequently, the effective azimuth resolution, $\Delta r_{\mathrm{azi,eff}}^{\mathrm{RMA}}$, can be obtained by being extended from $\Delta r_{\mathrm{azi}}^{\mathrm{RMA}}$ until the resultant $IRPR^{\mathrm{RMA}}$ equals to $IRPR^{\mathrm{ideal}}$. The effective range resolution, $\Delta r_{\mathrm{ran,eff}}^{\mathrm{RMA}}$ is assumed to be equal to $\Delta r_{\mathrm{ran}}^{\mathrm{ideal}}$.

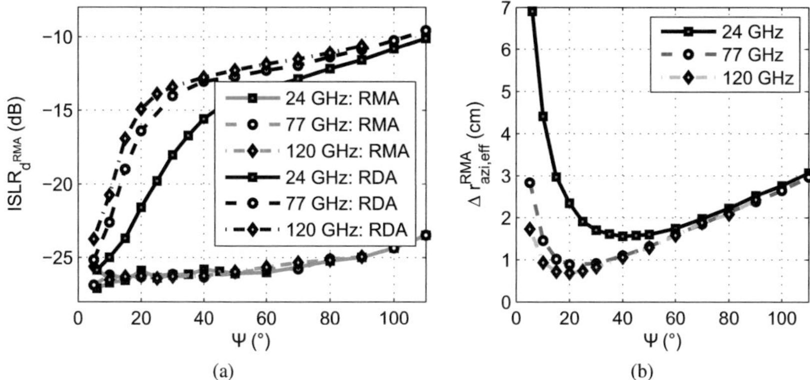

Figure 2.9: System performance as a function of Ψ: (a) $ISLR_{d_{azi}^{RMA}}^{RMA}$ and $ISLR_{d_{azi}^{RMA}}^{RDA}$, (b) $\Delta r_{azi,eff}^{RMA}$. SAR systems with $f_c = 24\,\text{GHz}$, $f_c = 77\,\text{GHz}$ and $f_c = 120\,\text{GHz}$ are simulated respectively with $B = 1\,\text{GHz}$.

4. As depicted in Figure 2.7d, the IRF of RDA is primarily distorted in azimuth direction. In principle, an extended mainlobe can be determined and consequently the effective resolution of RDA can be determined in a similar way.

The performances of RDA and RMA are evaluated using the metric presented above with the SAR simulator implemented in MATLAB, the results are shown in Figure 2.9.

As shown in Figure 2.9a, $ISLR_{d_{azi}^{RMA}}^{RDA}$ measured using d_{azi}^{RMA} is obviously larger than $ISLR_{d_{azi}^{RMA}}^{RMA}$ for all f_c when $\Psi > 10°$, and the difference between $ISLR_{d_{azi}^{RMA}}^{RDA}$ and $ISLR_{d_{azi}^{RMA}}^{RMA}$ is proportional to Ψ. In actual, even extending both d_{ran} and d_{azi} cannot decrease the resultant $ISLR^{RDA}$ to $ISLR_{d_{ideal}}^{ideal}$ when $\Psi > 20°$. Therefore it makes no sense to estimate $\Delta r_{ran,eff}^{RDA}$ and $\Delta r_{azi,eff}^{RDA}$ via $IRPR^{RDA}$. Since the deterioration of system performance in RDA for wide Ψ is unacceptable, RMA should be used for processing the SAR data.

Theoretically, as (2.25) indicates, Δr_{azi} is inversely proportional to Ψ. However, the broadening effect of the mainlobe is proportional to the Ψ. As shown in Figure 2.9b, there is an optimum Ψ for RMA in the sense of obtaining the minimum Δr_{azi}. For $f_c = 24\,\text{GHz}$ the minimum $\Delta r_{azi,eff}^{RMA} = 1.56\,\text{cm}$ can be achieved when $\Psi = 40°$; for

Table 2.1: Comparison of two metrics for adaptability of Ψ, with $B = 1\,\mathrm{GHz}$

f_c (GHz)	24		77		120	
Metric Type	A	B	A	B	A	B
Ψ (°)	40.00	19.41	20.00	10.72	20.00	8.57
Δr_{ran} (cm)	17.85	26.71	17.85	27.76	17.85	26.76
Δr_{azi} (cm)	1.56	2.22	0.89	1.24	0.69	1.00
$\Delta r_{\mathrm{ran}} \cdot \Delta r_{\mathrm{azi}}$ (cm^2)	27.85	59.33	15.89	33.28	12.32	26.68

$f_c = 77\,\mathrm{GHz}$ the minimum $\Delta r_{\mathrm{azi,eff}}^{\mathrm{RMA}} = 0.89\,\mathrm{cm}$ and for $f_c = 120\,\mathrm{GHz}$ the minimum $\Delta r_{\mathrm{azi,eff}}^{\mathrm{RMA}} = 0.69\,\mathrm{cm}$ can be achieved both when $\Psi = 20°$.

Another way to process SAR data with wide Ψ is to extract a part of fully filled rectangular support band from the original annular support band to generate the SAR image [113, 52]. In [113] a metric has been presented, based on which optimum Δr_{ran}, Δr_{azi} and Ψ are derived analytically, given f_{min} and B of the chirp. In principle, these optimum values are obtained when $\Delta r_{\mathrm{ran}} \cdot \Delta r_{\mathrm{azi}}$ is minimum, by selecting a rectangular support band of the maximum area from the whole support band. Therefore the bandwidth of k_y is inevitably smaller than the nominal bandwidth dictated by B, i.e. in this case Δr_{ran} is worse than $\Delta r_{\mathrm{ran}}^{\mathrm{ideal}}$. For comparison, the optimum Ψ, Δr_{ran} and Δr_{azi} for different f_c are listed in Table 2.1, where A is the metric of effective resolution presented in this dissertation, and B is the metric of minimum area resolution as presented in [113]. It is obvious that the resolutions obtained via metric A are better than the ones obtained via metric B, while $\Delta r_{\mathrm{ran,eff}}^{\mathrm{RMA}}$ is especially reserved to the theoretical $\Delta r_{\mathrm{ran}}^{\mathrm{ideal}}$.

The optimum Ψ obtained via the metric does not restrict the physical dimension of antenna, and the real Ψ can be artificially reduced by an azimuth window applied on the spatial frequency. However, as will be investigated in Chapter 3, the influence of some motion errors is proportional to Ψ, which is transferred from x_{am} domain via azimuth FFT to the whole k_x domain, and therefore cannot be alleviated by the azimuth frequency window.

It should be noted that the optimum Ψ obtained by the presented metric not only depends on f_c but also on B of the chirp and the window applied on the spatial frequency domain. For a specific application, a specific procedure of the metric should be applied with the given parameters.

3 Influences of Motion Errors

In an ideal case, the phase term of the SAR data in the spatial frequency domain, (2.52), changes linearly over the 2D support band. A 2D IFFT of (2.52) produces a IRF with a resolution determined by the shape of the support band. In practice, the motion errors deviate the radar sensor from the nominal straight trajectory and introduce undesirable components in the raw SAR data, which are subsequently transferred to the 2D spectrum of the SAR data after azimuth Fourier transform. Depending on their natures and magnitudes, the motion errors may cause the deterioration of SAR imaging in the following ways:

- Position error in range;

- Broadening of the range IRF, i.e. increase of $\Delta r_{\mathrm{ran,eff}}$;

- Position error in azimuth;

- Broadening of the azimuth IRF, i.e. increase of $\Delta r_{\mathrm{azi,eff}}$;

- Increase of the *ISLR*;

- Decrease of *SNR*.

Knowledge of the influences of motion errors and their allowable levels is required to determine whether motion compensation is necessary for near-range applications, and to determine the parameters of the motion compensation algorithms if necessary.

A comprehensive investigation is presented in this chapter, which starts from the motion-error-deteriorated raw SAR data, instead of analysis based on phase errors in the spatial frequency domain as proposed in [18]. As results of the near-range geometry and wide antenna beamwidth, two specific aspects are exclusively covered in this dissertation. First, the investigated motion errors parameters are designed according to the degree of discomfort (DoD) defined in [114], which is an ISO standard for evaluating the human's comfort in the public transport. Second, some motion errors, which need to be compensated for far range targets, are artificially introduced into the relatively near range targets. Such motion errors are called artificial motion errors in this dissertation and their influences on near range targets are quantified and a special motion compensation algorithm is designed accordingly in Chapter 4.

The definition and classification of motion errors are given in section 3.1, where the nominal straight trajectory with which the motion error measurements are compared to is defined for near-range SAR applications. In section 3.2 the influences of translation motion errors have been investigated and their allowable levels have been obtained by analysis, as well as by computer simulation. The analysis of translation motion errors along the azimuth direction is divided into initial position error, constant velocity, constant acceleration and higher order vibrations in section 3.3. In section 3.4, the analysis based on the same classification is performed for the range motion errors. The comparability of vertical motion errors to range motion errors is given in section 3.5, where it is shown that the vertical motion errors cause similar distortions on the SAR image, but with much less amplitude.

There are two methods in terms of the generation of raw SAR data. The time-domain method first generates the SAR data of each target and then sums them to form the final raw data of the whole scene. On the contrary, the frequency-domain method generates the raw SAR data via a superposition integral in which the whole scene is weighted by the IRF of the SAR system [115, 116, 117]. In order to avoid effects other than that caused by motion errors, the generation of raw SAR data containing motion errors uses the time-domain method, instead of the more efficient but less accurate frequency-domain method.

3.1 Definition and Classification of Motion Errors

In this dissertation, the radar sensors, i.e. the transmitting and receiving antennas, are assumed to be firmly mounted on the vehicle, therefore the radar-vehicle is considered as a rigid body. Since no servo for the antenna is applied, the center of the beam always points to a constant Doppler frequency, and for simplicity and without loss of generality zero Doppler is assumed, i.e. the squint angle of antenna is zero. The grazing angle of the radar, i.e. the angle formed by the beam center and the ground,

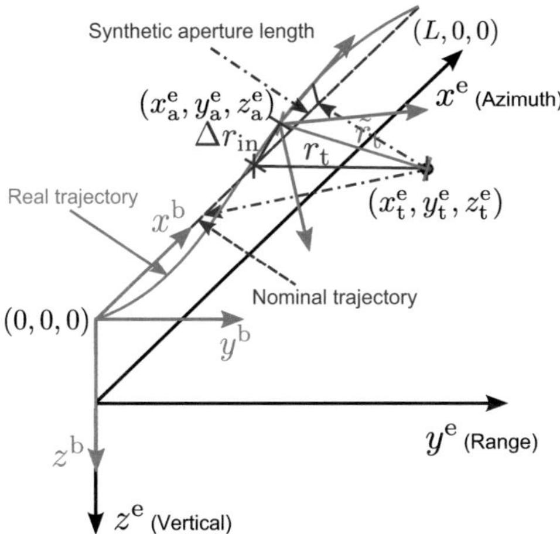

Figure 3.1: Geometry of SAR expressed in Earth-fixed frame.

is also assumed to be zero in this chapter for the ease of explanation, while the more general cases with non-zero grazing angle are discussed in Chapter 4.

3.1.1 Definition

The geometry of the near-range SAR imaging in Earth-fixed frame is illustrated in Figure 3.1. The SAR processing is virtually performed in Earth-fixed frame (denoted with superscript "e") with its axes aligned with the local vertical and a plane which is tangential to the Earth's surface. The body frame denoted with superscript "b" is where the inertial measurement unit (IMU) measures the instantaneous acceleration and angular rates of the radar sensors, whereby the coordinates of the radar sensors in Earth-fixed frame are calculated. The relationship between Earth-fixed frame and body frame will be discussed in detail in Chapter 4. The final SAR image is presented in the image display plane (IDP). It is the plane formed by the nominal antenna beam center line on which the targets along the line-of-sight (LOS) of the radar are projected. Since the grazing angle is assumed to be zero in this chapter, IDP is actually the $x^e y^e$-plane.

Two trajectories - the real trajectory and the nominal trajectory - in one block length L are illustrated in Figure 3.1, where L is the basic unit of SAR processing as depicted in Figure 2.2. The raw SAR data are collected along the real trajectory of the radar sensor, while the nominal trajectory is artificially defined based on the measurements of the real trajectory, generally through a least-squares fit. Once the nominal trajectory is determined, the directions of x^e and ye axes of the Earth-fixed frame are defined, where x^e axis is along the nominal velocity and y^e axis is perpendicular to the plane formed by x^e and z^e axes. In an ideal case, the radar sensor traverses a straight line path along the nominal axis x^e at a constant velocity v, which coordinates are $(vt, 0, 0)$, and the antenna is constantly directed at the broadside without any rotation. However, motion errors always exist in real situation, the real trajectory of the radar sensor, which coordinates are (x_a^e, y_a^e, z_a^e), deviates from the nominal trajectory as a combination of translation and rotation errors.

On the one hand, the motion errors with regard to the nominal trajectory can be expressed as

$$\Delta r_{in} = \sqrt{(x_a^e - vt)^2 + (y_a^e)^2 + (z_a^e)^2}, \tag{3.1}$$

where the subscript "$_{in}$" denotes that such motion errors can be considered as the space-invariant component of the motion errors for all targets. The term "motion errors" refers to Δr_{in} in this dissertation unless otherwise stated. On the other hand, the slant range between the antenna and a certain target located at (x_t^e, y_t^e, z_t^e) changes from r_t to $\tilde{r}_t$ as a function of the antenna's instantaneous coordinates (x_a^e, y_a^e, z_a^e). The difference between r_t and $\tilde{r}_t$ is called the motion errors for target t, Δr_t, which can be expressed as

$$\begin{aligned} \Delta r_t &= \tilde{r}_t - r_t \\ &= \sqrt{(x_a^e - x_t^e)^2 + (y_a^e - y_t^e)^2 + (z_a^e - z_t^e)^2} - \sqrt{(vt - x_t^e)^2 + (y_t^e)^2 + (z_t^e)^2}. \end{aligned} \tag{3.2}$$

In addition, the difference between Δr_t and Δr_{in} is the space-variant component of the motion errors

$$\Delta r_{var} = \Delta r_t - \Delta r_{in}. \tag{3.3}$$

It is clear that Δr_{var} depends on the target's location and is generally non-zero. For the ease of explanation, a constant deviation from the nominal trajectory in y^e axis in the 2D $x^e y^e$-plane and different components of the motion errors are illustrated in

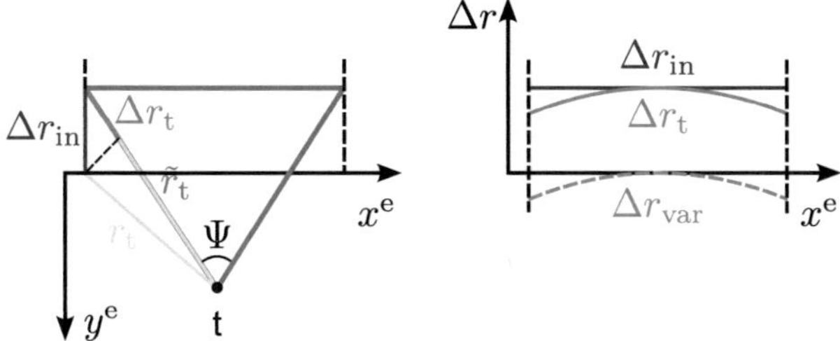

Figure 3.2: The relationship between Δr_t, Δr_{in} and Δr_{var}.

Figure 3.2. It can be seen that Δr_t is not linear when the real trajectory is linear, and this results from the fact that the range equation of a target, r_t, is hyperbolic as indicated in (2.8). When the antenna beamwidth is narrow, e.g. $\Psi < 10°$ in remote sensing SAR operating in C or X frequency band [97, 118], Δr_{var} can be ignored. However, as Δr_{var} grows with Ψ at the edge of the synthetic aperture, it becomes non-negligible when Ψ is larger than a certain value. For instance, the antenna beamwidth can be larger than $30°$ in a SAR system working at VHF/UHF band [52]. Moreover for a near-range SAR system which intends to take advantage of the already existing automotive radar, Ψ can easily be larger than $60°$ [25, 32]. In addition, it is intuitive that the absolute value of Δr_{var} is inversely proportional to the target's range of the closest approach. As result, after the first-order compensation for Δr_{in}, which is the common step of most motion compensation algorithms [119], the maximum absolute value of residual Δr_{var} of the near range targets is larger than that of the far range targets.

The rotation of the radar puts irregular amplitude modulation (AM) on the SAR raw data via the antenna pattern. In this dissertation an antenna pattern of sinc function both in azimuth with a HPBW of Ψ and elevation with a HPBW of Φ is used, as illustrated in Figure 3.3. Without rotation errors, the normal AM for a certain target related to antenna pattern is the contour of the slice in parallel with the $x^e y^e$-plane. Rotation errors cause the regularly shaped AM to become an irregular route on the pattern surface. The severe irregular AM can cause azimuth aliasing, radiometric loss via antenna beam pattern and depress the SNR of the image [120, 121].

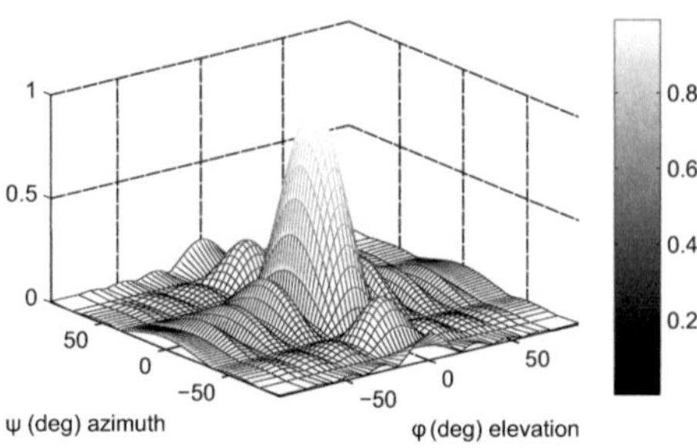

Figure 3.3: Antenna amplitude pattern with $\Psi = 33°$ and $\Phi = 17°$.

Deteriorated by the translation and rotation errors, the SAR raw IF signal can now be expressed as

$$
\begin{aligned}
\tilde{s}_{\mathrm{IF}}(x_{\mathrm{am}}, k_r) &= A_t w_r(k_r) w_a(\tilde{r}_t) \exp(-jk_r\tilde{r}_t) \\
&= A_t w_r(k_r) w_a(r_t) \exp(-jk_r r_t) \\
&\quad \cdot \frac{w_a(\tilde{r}_t)}{w_a(r_t)} \cdot \exp(-jk_r\Delta r_t) \\
&= s_{\mathrm{IF}}(x_{\mathrm{am}}, k_r) \cdot \frac{w_a(\tilde{r}_t)}{w_a(r_t)} \cdot \exp(-jk_r\Delta r_t).
\end{aligned}
\tag{3.4}
$$

In (3.4), the translation errors for a certain target t, Δr_t, introduces a phase term, $\exp(\cdot)$ in the raw SAR data, while the rotation errors of the sensor which change the LOS of the radar put a amplitude error term, $\frac{w_a(\tilde{r}_t)}{w_a(r_t)}$, on the raw SAR data. In the sake of brevity, the superscript "^{e}" denoting Earth-fixed frame and the subscript "$_a$" denoting antenna are omitted in the following equations within this chapter.

Provided that the fluctuations of motion errors are smaller than that of the motion-error-free SAR signal, $s_{\mathrm{IF}}(x_{\mathrm{am}}, k_r)$, which is a chirp signal along both x_{am} and k_r directions as described in Chapter 2, the motion error components can be modeled by $\frac{w_a(\tilde{r}_t)}{w_a(r_t)} \cdot \exp(-jk_r\Delta r_t)$. This term can then be modeled approximately as a phase error term in the spatial frequency domain (k_x, k_r) with a substitution of the variable by (2.47) according to POSP. This is the prerequisite of the analytical method for investigating motion errors since there is a linear mapping relationship between the motion errors in space domain and the spatial frequency domain. However, if the fluctuations of motion errors are comparable to that of the motion-error-free SAR signal, an analytical expression of the SAR data in the 2D spectrum cannot be obtained, which contains both amplitude and phase errors caused by the motion errors. In this case, a computer simulation, which starts from (3.4) instead of the expression in 2D spectrum, is the only way to investigate the influences of motion errors. In the following sections, both methods are applied during the analysis according to the frequencies of the motion errors.

3.1.2 Classification

According to their influences on the SAR image, motion errors with regard to the nominal trajectory, Δr_{in}, can be classified into three types: linear, quadratic and sinusoidal [18]. As depicted in Figure 3.4, the horizontal axis is the nominal velocity multiplied by time in unit of m and the vertical axis is the amplitude of the motion errors, for translation errors in unit of m and for rotation errors in unit of rad or $°$. The classification depends on the target's synthetic aperture length, L_t. The same sinusoidal motion error can be a linear error over the short enough synthetic aperture L_1, or a quadratic error over the synthetic aperture L_2, or a sinusoidal error with one or multiple cycles over the long synthetic aperture L_3. Since the geometry is represented in three-dimensional (3D) space instead of 2D space as depicted in Figure 2.2, the synthetic aperture length of a certain target is now expressed as

$$L_t = 2 \cdot \sqrt{y_t^2 + z_t^2} \cdot \tan\left(\frac{\Psi}{2}\right), \tag{3.5}$$

where $\sqrt{y_t^2 + z_t^2}$ is the range of the closest approach, $r_{t,0}$, in the 3D space, which is also the coordinate y of the target in the final 2D SAR image, i.e. in IDP.

It is seen from Figure 3.4 that the duration of a target being illuminated by the radar along azimuth direction is also a function of the nominal velocity, v. A parameter called motion frequency to synthetic aperture frequency ratio, MSR, to classify motion errors has been defined, which is similar to the one defined in [18] and can be expressed as

$$MSR = \frac{f_m}{f_{\text{syn}}} = \frac{f_m}{\frac{v}{L_t}} = \frac{L_t}{\frac{v}{f_m}} = \begin{cases} [0, 0.25] & \text{Linear} \\ [0.25, 0.75] & \text{Quadratic} \\ [0.75, \infty] & \text{Sinusoidal} \end{cases} . \tag{3.6}$$

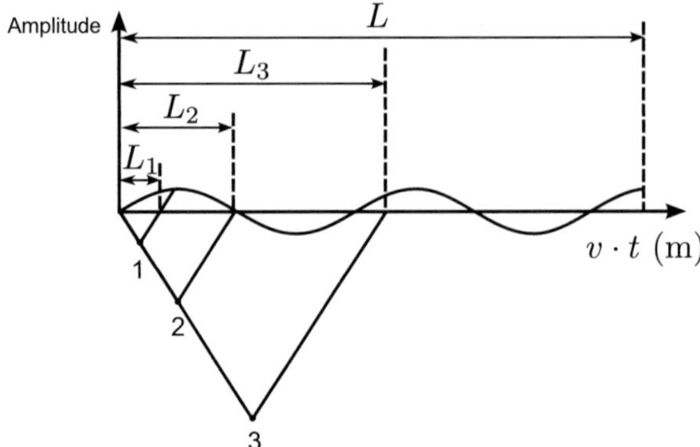

Figure 3.4: Classification of motion errors with regard to synthetic aperture length: linear for target 1, quadratic for target 2 and sinusoidal for target 3.

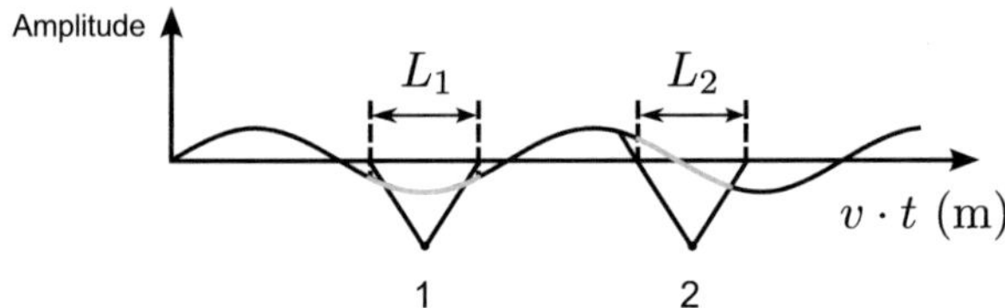

Figure 3.5: Dependence on initial position of the synthetic aperture of classification of motion errors.

In (3.6), f_m is the frequency of the motion errors, and f_syn is the synthetic aperture frequency, which is the reciprocal of synthetic aperture time, $\frac{L_\mathrm{t}}{v}$. Note that the term $\frac{v}{f_\mathrm{m}}$ is the period of the motion error in unit of m, and therefore MSR is also the ratio between the target's synthetic aperture length and the motion error's period. According to (3.6), if L_t covers more than a quarter of the motion error's period, the motion error is quadratic or sinusoidal to the target. However, as depicted in Figure 3.5, a motion error with $MSR > 0.25$ can still be considered as linear within the synthetic aperture of target 2, while within the synthetic aperture of target 1 the same motion error with the same MSR is obviously quadratic. So the classification of motion errors is actually also related to the initial position of the synthetic aperture with respect to the motion error. The most strict definition given in (3.6) intends to cope with adverse conditions.

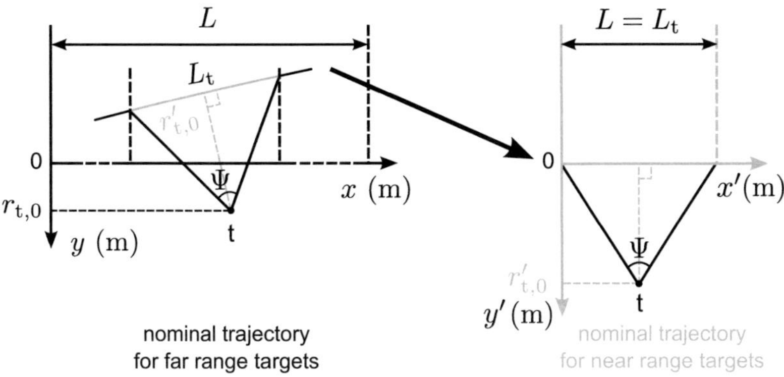

Figure 3.6: The loss of relative position information caused by artificial motion errors.

3.1.3 Artificial Motion Errors

The concept of artificial motion errors is specifically related to translation errors in range and vertical directions. For near-range applications, the nominal trajectory defined according to the furthest targets could be very different from the nominal trajectories defined according to the nearest targets. Under the extremist conditions, the real trajectory which contains the motion errors with regard to the furthest targets can be considered as the conjunction of a series of short nominal trajectories with regard to the nearest targets. In such a case, the relative position between the antenna and the nearest targets is deteriorated by the motion compensation. Therefore such motion errors as part of the quadratic or higher-order motion errors defined with regard to the long-term nominal trajectory for far range targets while being linear within the synthetic aperture of near range targets are called artificial motion errors in this dissertation.

For instance, in Figure 3.6, the same target's location in IDP with and without motion compensation, i.e. with regard to nominal trajectory defined for far range targets and defined for near range targets, has been illustrated in the left and right side respectively. Note that the yaw angle (rotation) between the nominal and real trajectory is illustrated, and the SAR imaging geometry is presented in a 2D plane instead of a 3D space for the ease of explanation. The primary component of motion errors within the synthetic aperture of target t (the gray part of the real trajectory) is introduced artificially as a result of defining the nominal trajectory according to targets located at the further range region. If the length of the SAR processing block, L, reduces to L_t, this

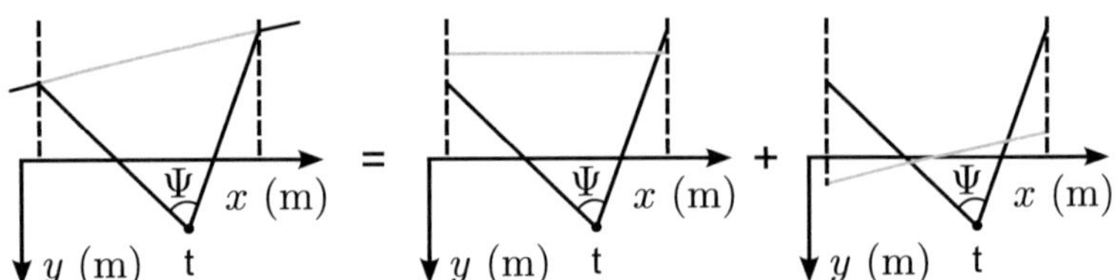

Figure 3.7: Decomposition of a linear artificial motion error.

short-term nominal trajectory would almost overlap with the real trajectory. By registering to this new short-term nominal trajectory, motion errors reduce almost to zero. In the following discussion, such coordinate system dictated by the real short-term trajectory with regard to a certain target is called $x'y'$-coordinate system and denoted with the superscript "'", as opposed to the xy-coordinate system dictated by the nominal long-term trajectory. For near range target t, the real range of the closest approach is $r'_{t,0}$ in $x'y'$-coordinate system. However, after applying motion compensation for Δr_{in} in xy-coordinate system this value changes to $r_{t,0}$.

Generally, the first step of motion compensation is to compensate the space-invariant component of the motion errors. From (3.2), (3.1) and (3.3), the expression of the IF data changes from (3.4) to

$$\tilde{s}_{\text{IF,in}}\left(x_{\text{am}}, k_r\right) = \tilde{s}_{\text{IF}}\left(x_{\text{am}}, k_r\right) \cdot \exp\left(jk_r \cdot \Delta r_{\text{in}}\right)$$

$$= s_{\text{IF}}\left(x_{\text{am}}, k_r\right) \cdot \frac{w_{\text{a}}\left(\tilde{r}_t\right)}{w_{\text{a}}\left(r_t\right)} \cdot \exp\left(-jk_r \cdot \Delta r_{\text{var}}\right). \tag{3.7}$$

At this step, for near range targets, the relative position information is lost, which in principle can still be recovered later from the measured trajectory and the final SAR image. If an antenna of wide Ψ is used and no further motion compensation is applied, Δr_{var} can be substantial at the edge of the synthetic aperture, especially when a great linear motion error deviated from the nominal trajectory exists. As explained in subsection 3.1.1, Δr_{var} of near range targets are higher than that of far range targets. Therefore, besides the loss of relative position information, artificial motion errors also deteriorate the final SAR image in the near range region by introducing a larger Δr_{var} if no further motion compensation is applied.

As depicted in Figure 3.7, the artificial motion errors are linear within the synthetic aperture of a certain target and can always be decomposed into a constant deviation

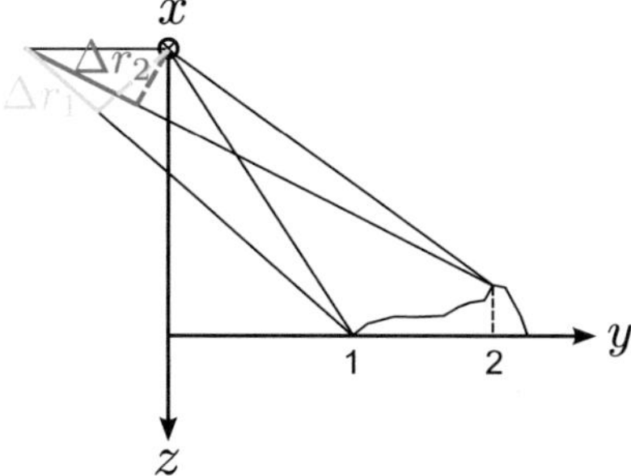

Figure 3.8: Range-dependent motion errors in remote sensing applications.

in parallel with the nominal trajectory and a linear deviation trajectory. Here the 2D representation is used for simplicity and without loss of generality. According to (3.7), the effects of Δr_{var} caused by these two types of Δr_{in} will be analyzed in subsection 3.4.1 and 3.4.2 respectively.

The residual motion errors after first-order motion compensation, i.e. Δr_{var}, are space-variant, and the second-order motion compensation to address this problem is usually computationally expensive [104, 102, 105, 119, 108]. In Chapter 4, an alternative motion compensation algorithm is proposed to solve this problem by exploiting the feature of the geometry of near-range applications.

It should be noted that the artificial motion errors are different from the range-dependent motion errors in remote sensing applications, which are caused by the different incidence angles of range bins as well as the different heights of the targets. As depicted in Figure 3.8, the difference between Δr_1 and Δr_2 is a function of the ranges and heights of targets. To deal with these differences a second-order motion compensation using digital elevation model (DEM) is required [97, 48, 60, 105, 119]. On the contrary, as illustrated in Figure 2.1b, in near-range applications the targets at the same height level with the radar are of interest. Therefore such range dependence is not considered further in this dissertation. In addition, the motion compensation for linear motion errors in remote sensing applications is to register all the targets to a long common trajectory, and hence the absolute position information is more important than the relative position between the plane and the targets, which is usually not the case for near-range automotive applications.

3.2 Vehicle's Dynamics Model

To investigate the influences of motion errors, the dynamics of the vehicle-radar rigid body is required. For a nominal trajectory of SAR, the vehicle's dynamics model in Earth-fixed frame can be expressed as

$$
\begin{cases}
x(t) = v \cdot t \\
y(t) = 0 \\
z(t) = 0
\end{cases}
, \tag{3.8}
$$

where v is the nominal constant velocity.

However, in a real situation, the velocity of a vehicle as a vector is always changing in the azimuth, range and vertical direction, as a result of the intentional maneuver, environment turbulence or the vehicle's inherent vibration. In these cases the dynamics model of the vehicle becomes

$$
\begin{cases}
\tilde{x}(t) = v \cdot t + \Delta x(t) \\
\tilde{y}(t) = \Delta y(t) \\
\tilde{z}(t) = \Delta z(t)
\end{cases}
, \tag{3.9}
$$

where $\Delta x(t)$, $\Delta y(t)$ and $\Delta z(t)$ are motion errors with respect to the nominal trajectory, and $t \in [0, t(L)]$ with $t(L)$ being the length of the total SAR processing in time unit. According to the classification defined in section 3.1, within a certain target's synthetic aperture of length L_t, motion errors along each axis incorporate an initial position error, a constant velocity error, $\Delta v \cdot t$, a constant acceleration error, $\frac{1}{2}\Delta a \cdot t$, and the sum of other high sinusoidal frequency errors, $\sum_i A_i \cdot \sin(2\pi f_i t)$, which can be expressed as

$$
\begin{cases}
\Delta x(t) = \Delta x_0 + \Delta v_x t + \frac{1}{2}\Delta a_x \cdot t^2 + \sum_i A_{xi} \cdot \sin(2\pi f_{xi} t) \\
\Delta y(t) = \Delta y_0 + \Delta v_y t + \frac{1}{2}\Delta a_y \cdot t^2 + \sum_i A_{yi} \cdot \sin(2\pi f_{yi} t) \\
\Delta z(t) = \Delta z_0 + \Delta v_z t + \frac{1}{2}\Delta a_z \cdot t^2 + \sum_i A_{zi} \cdot \sin(2\pi f_{yi} t)
\end{cases}
, \tag{3.10}
$$

where $t \in [0, t(L_t)]$ with $t(L_t)$ being the length of synthetic aperture of target t in time unit.

For airborne SAR where the antenna is pointed to a fixed direction via antenna servo, the investigation of the residual rotation errors is generally coupled with translation errors, i.e. the parameters of rotation errors are related to the translation motion errors. For example, the aerodynamic coupling between translation and rotation has been investigated in [54] for motion errors caused by typical gust. Also, in [55] the rotation errors either caused by a weather disturbance or a intentional maneuver were assumed to be coupled with the translation errors. Since the scheme without antenna servo is assumed in this dissertation, the antenna is always pointed to the zero Doppler frequency or a certain Doppler value. As result, rotations about x, y and z axis, i.e. roll, pitch and yaw, are coupled with translations in a deeper way than that in the airborne SAR cases. Therefore, the investigations of the influences of rotation errors, represented with the amplitude modulation term $\frac{w_a(\tilde{r}_t)}{w_a(r_t)}$, are performed with translation errors in section 3.3, 3.4 and 3.5 respectively.

3.3 Azimuth

The azimuth motion errors cause the spatial sampling of SAR along azimuth direction to become uneven or deviate from its nominal value, thereby introducing defocus and distortion in the final image. The analysis of the influences of Δv_x and Δa_x in (3.10) are performed analytically with the tool provided in [19], i.e. separating the Stolt mapping into three components: RCMC, azimuth compression and range-azimuth decoupling. However, the analytical deductions are completely in the spatial frequency domain and without approximation of a narrow beamwidth. On the contrary, the influence of higher order sinusoidal components, $\sum_i A_{xi} \cdot \sin\left(2\pi f_{xi} t\right)$, is investigated experimentally using the SAR simulator with the motion parameters designed according to ISO 2631 as described in Appendix A.

3.3.1 Initial Position Error

Mathematically, Δx_0 results from the preceding azimuth errors, it registers the target ahead or behind its nominal position in the azimuth direction with Δx_0. In practice the azimuth motion errors are usually compensated by controlling the *PRF* according to the instantaneous azimuth velocity or by resampling the unevenly sampled raw data along azimuth direction. As result, Δx_0 is compensated with other components of the azimuth motion errors, and the target is registered to the correct position.

3.3.2 Constant Velocity Error

Considering that only a constant velocity error, Δv_x, exists, the sensor's instantaneous velocity in azimuth direction can be expressed as

$$\tilde{v}_x = \frac{d\tilde{x}(t)}{dt} = v + \Delta v_x. \tag{3.11}$$

The motion errors are transferred to the spatial frequency k_x via performing azimuth FFT on (3.4), the deteriorated SAR signal represented in (k_x, k_r) domain is

$$\begin{aligned}
\tilde{S}_{\mathrm{IF,nsg}}\left(\tilde{k}_x, k_r\right) &= \tilde{S}_{\mathrm{IF}}\left(\tilde{k}_x, k_r\right) \cdot \exp\left(-j\tilde{k}_x \cdot \tilde{v}_x \hat{t}\right) \\
&= A_1 w_r\left(k_r\right) W_a\left(\tilde{k}_x, k_y\right) \exp\left[-j\left(\tilde{k}_x \cdot x_t + \sqrt{k_r^2 - \tilde{k}_x^2} \cdot y_t\right)\right].
\end{aligned} \tag{3.12}$$

Note that $\tilde{v}_x$ does not affect the "non-stop-and-go" compensation, since $\tilde{k}_x$ is inversely proportional to $\tilde{v}_x$ as dictated in (2.31).

In addition, combining (2.18) and (2.31), k_x can be expressed as

$$k_x(v) = k_r \cdot \sin(\psi) = k_r \cdot \frac{vt}{\sqrt{y_t^2 + (vt)^2}}. \tag{3.13}$$

and the derivative of k_x with respect to the nominal velocity, v, can be calculated as

$$\frac{dk_x}{dv} = \frac{k_r}{v} \sin(\psi) \cos^2(\psi). \tag{3.14}$$

So the effect of Δv_x in space domain is introduced to the spatial frequency domain via (3.14).

First, the errors in $-\tilde{k}_x \cdot x_{\mathrm{t}}$ caused by Δv_x, in the case of small Ψ, can be approximated as

$$\begin{aligned}
\Delta\phi_{\mathrm{azishift}} &= -x_{\mathrm{t}} \cdot \frac{dk_x}{dv} \Delta v_x \\
&\approx -x_{\mathrm{t}} \cdot \frac{\Delta v_x}{v} \cdot k_x.
\end{aligned}$$
(3.15)

The effect of (3.15) is to shift the target in the final SAR image along the x axis in accordance with the following equation:

$$\tilde{x}_{\mathrm{t}} = x_{\mathrm{t}} + x_{\mathrm{t}} \cdot \frac{\Delta v_x}{v}.$$
(3.16)

It is clear that $\frac{\Delta v_x}{v}$ changes the scale of the nominal x axis linearly.

Second, since the Stolt mapping, as a change of variables $k_y = \sqrt{k_r^2 - k_x^2}$, performs the RCMC, azimuth compression and range-azimuth decoupling together, the errors in $-\sqrt{k_r^2 - \tilde{k}_x^2} \cdot y_{\mathrm{t}}$ caused by Δv_x have influences on all these three processes. To investigate these effects separately, $-\sqrt{k_r^2 - \tilde{k}_x^2} \cdot y_{\mathrm{t}}$ can be expanded as a power series up to quadratic terms in k_r and $\tilde{k}_x$, as deduced in [19], as

$$-\sqrt{k_r^2 - \tilde{k}_x^2} \cdot y_{\mathrm{t}} \approx -y_{\mathrm{t}} \cdot \left[k_{rc}D + \frac{k_{rr}}{D} + \frac{k_{rr}^2 \cdot \tilde{k}_x^2}{2k_{rc}^3 D^3} \right],$$
(3.17)

with

$$D = \sqrt{1 - \left(\frac{\tilde{k}_x}{k_{rc}} \right)^2},$$
(3.18)

$$k_{rc} = \frac{4\pi}{c} \cdot f_{\mathrm{c}},$$
(3.19)

$$k_{rr} = \frac{4\pi}{c} \cdot f_r.$$
(3.20)

The first term in the square brackets represents the azimuth compression, the second term represents the RCMC and the third term represents the range-azimuth decoupling.

To consider the azimuth compression, note that the first term can be approximated as

$$-y_t k_{rc} D \approx -y_t k_{rc} + \frac{y_t}{2k_{rc}} \tilde{k}_x^2. \qquad (3.21)$$

By comparing (3.21) to (2.39), it is seen that the term $\frac{y_t}{2k_{rc}} \tilde{k}_x^2$ performs the azimuth compression of RMA. The velocity error in $\tilde{k}$ introduces phase errors in the azimuth matched filter as

$$\Delta\phi_{\text{azicom}} = \frac{y_t}{2k_{rc}} 2k_x \frac{dk_x}{dv} \Delta v_x. \qquad (3.22)$$

By substituting (3.14) and (3.13) into (3.22), $\Delta\phi_{\text{azicom}}$ can be expressed as

$$\begin{aligned}
\Delta\phi_{\text{azicom}} &= \frac{k_r^2}{4k_{rc}} \cdot y_t \cdot \frac{\Delta v_x}{v} \cdot \sin^2(2\psi) \\
&= \frac{y_t}{k_{rc}} \cdot \frac{\Delta v_x}{v} \cdot k_x^2 \cdot \cos^2(\psi),
\end{aligned} \qquad (3.23)$$

which is a quasi-quadratic phase error (QPE) in the spatial frequency domain. When Ψ is small, i.e. $\cos(\psi) \approx 1$, (3.23) can be approximated to a standard quadratic function of k_x as

$$\Delta\phi_{\text{azicom}} \approx \frac{y_t}{k_{rc}} \cdot \frac{\Delta v_x}{v} \cdot k_x^2. \qquad (3.24)$$

As suggested in [18], multiplication by a QPE term in the spatial frequency domain introduces a defocus in the space domain. The "width" of the defocus in space domain, x_{QPE}, can be approximated as

$$\begin{aligned}
x_{\text{QPE}} &= 2 \cdot \frac{y_t}{k_{rc}} \cdot \frac{\Delta v_x}{v} \cdot B_{k_x} \\
&= 2 \cdot \frac{y_t}{\frac{4\pi}{c} \cdot f_c} \cdot \frac{\Delta v_x}{v} \cdot \frac{8\pi}{\lambda} \sin\left(\frac{\Psi}{2}\right) \\
&\approx 4 y_t \sin\left(\frac{\Psi}{2}\right) \cdot \frac{\Delta v_x}{v},
\end{aligned} \qquad (3.25)$$

which is proportional to y_t, Ψ and $\frac{\Delta v_x}{v}$, but independent of f_c.

The second term in (3.17) corresponds to RCMC and can be expanded up to terms in k_x^2 as

$$-y_t \frac{k_{rr}}{D} \approx -k_{rr}y_t - \frac{k_{rr}\tilde{k}_x^2}{2k_{rc}^2}y_t$$

$$= -2\pi f_r \cdot \frac{2\left(y_t + \frac{\lambda_c^2 y_t}{32\pi^2}\tilde{k}_x^2\right)}{c}. \tag{3.26}$$

By comparing (3.26) with (2.38), it is seen that $\frac{k_{rr}\tilde{k}_x^2}{2k_{rc}^2}y_t$ performs the RCMC of SAR data in the spatial frequency domain. The phase errors related to $\frac{k_{rr}\tilde{k}_x^2}{2k_{rc}^2}y_t$ caused by Δv_x can be calculated as

$$\Delta\phi_{\text{RCMC}} = -\frac{k_{rr}}{k_{rc}} \cdot \frac{y_t}{2k_{rc}} 2k_x \frac{dk_x}{dv}\Delta v_x$$

$$= -\frac{k_{rr}}{k_{rc}} \cdot \Delta\phi_{\text{azicom}}. \tag{3.27}$$

Since $\frac{k_{rr}}{k_{rc}}$ is independent of k_x, $\Delta\phi_{\text{RCMC}}$ also causes azimuth defocus in the final SAR image. When the relative bandwidth, $\frac{B}{f_c}$, of the radar system is small, the influence related to RCMC can be ignored since

$$\left|\frac{\Delta\phi_{\text{RCMC}}}{\Delta\phi_{\text{azicom}}}\right| = \left|\frac{k_{rr}}{k_{rc}}\right| \leqslant \frac{B}{2f_c} \ll 1. \tag{3.28}$$

Since the third term in (3.17) corresponds to the range-azimuth decoupling, the phase errors related to this term can be calculated as

$$\Delta\phi_{\text{ran}-\text{azi}} \approx -\frac{k_{rr}^2}{k_{rc}^2} \cdot \frac{y_t}{2k_{rc}} 2k_x \frac{dk_x}{dv}\Delta v_x$$

$$= -\frac{k_{rr}^2}{k_{rc}^2} \cdot \Delta\phi_{\text{azicom}}$$

$$= -\frac{y_t}{k_{rc}^3} \cdot \frac{\Delta v_x}{v} \cdot k_{rr}^2 \cdot k_x^2, \tag{3.29}$$

where the approximation is due to assuming $D \approx 1$. On the one hand, the influence of $\Delta\phi_{\text{ran}-\text{azi}}$ on azimuth defocus is much smaller than $\Delta\phi_{\text{azicom}}$ by analogy with (3.27), and therefore it can be ignored. On the other hand, $\Delta\phi_{\text{ran}-\text{azi}}$ is also a quadratic function of

k_{rr}, which causes range defocus in the final SAR image after range Fourier transform. The defocus effect on the Δr_{ran} and Δr_{azi} introduced by the same Δv_x can be compared by the ratio of the quadratic coefficient of k_{rr}^2 in (3.29) to the quadratic coefficient of k_x^2 in (3.23), which is

$$\frac{\frac{d\Delta\phi_{\text{ran-azi}}}{dk_{rr}^2}}{\frac{d\Delta\phi_{\text{azicom}}}{dk_x^2}} = \frac{-\frac{y_t}{k_{rc}^3}\cdot\frac{\Delta v_x}{v}\cdot k_x^2}{\frac{y_t}{k_{rc}}\cdot\frac{\Delta v_x}{v}\cdot\cos^2(\psi)} = \left[\frac{k_r\sin(\psi)}{k_{rc}\cos(\psi)}\right]^2 \leqslant \left(1+\frac{B}{2f_c}\right)^2\cdot\tan^2\left(\frac{\Psi}{2}\right). \quad (3.30)$$

According to the discussion in Chapter 2, the optimum value of Ψ in terms of effective Δr_{azi} are $40°$ for $f_c = 24\,\text{GHz}$, and $20°$ for both $f_c = 77\,\text{GHz}$ and $f_c = 120\,\text{GHz}$, and the corresponding maximum ratios are 14%, 3% and 3% respectively. Furthermore, the ratio of the bandwidth of k_{rr}, $B_{k_{rr}}$, to B_{k_x} can be calculated as

$$\frac{B_{k_{rr}}}{B_{k_x}} = \frac{\frac{4\pi}{c}B}{\frac{8\pi}{\lambda}\sin\left(\frac{\Psi}{2}\right)} \approx \frac{B}{2f_c\sin\left(\frac{\Psi}{2}\right)}. \quad (3.31)$$

By comparing (2.24) and (2.25) with (3.31), it is seen that the ratio of $B_{k_{rr}}$ is B_{k_x} is also the ratio of Δr_{ran} to Δr_{azi}. For the SAR systems of interest in this dissertation, $B_{k_{rr}}$ is much smaller than B_{k_x}. According to (3.25), since the influence of QPE is proportional to its coefficient and the frequency bandwidth, the influence of $\Delta\phi_{\text{ran-azi}}$ on range defocus can be ignored.

Recapitulating, the influence of Δv_x introduces a defocus effect on the final SAR image, and the deterioration of Δr_{azi} is much more severe than that of Δr_{ran} for the systems with optimum Ψ. Moreover, since $\Delta\phi_{\text{azicom}}$ in (3.23) is azimuth-invariant but range-dependent, the effect of defocus in the final SAR image is proportional to the range coordinates.

A series of $\frac{\Delta v_x}{v}$ have been applied on the same set of motion-error-free SAR raw data generated at $f_c = 24\,\text{GHz}$, $y_t = 1\,\text{m}$ and $\Psi = 40°$. The difference between the deteriorated and original $ISLR_{d\text{RMA}}$, denoted with $\Delta ISLR_{d\text{RMA}}$, are plotted in Figure 3.9a, where the subscript "$_{d\text{RMA}}$" denotes that the boundary of the mainlobe is defined by d^{RMA} as described in subsection 2.4.3. In addition, the defocus effect is evaluated by the broadening factor of $\Delta r_{\text{azi,eff}}$, denoted with $F_{\Delta r_{\text{azi}}}^{\Delta v_x}$. It is the ratio of the $\Delta r_{\text{azi,eff}}$ determined when there are motion errors to the $\Delta r_{\text{azi,eff}}$ determined when there is no motion error. In Figure 3.9b, $F_{\Delta r_{\text{azi}}}^{\Delta v_x}$ is plotted as a function of $\frac{\Delta v_x}{v}$.

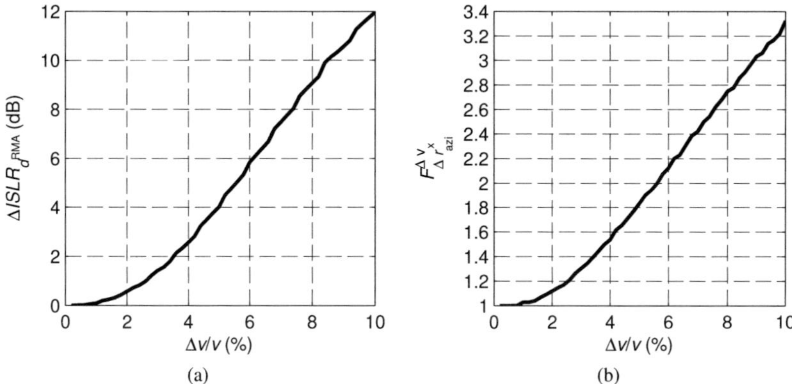

Figure 3.9: The influences of Δv_x in the case of $f_c = 24\,\text{GHz}$, $y_t = 1\,\text{m}$ and $\Psi = 40°$: (a) increasing of $ISLR_{d\text{RMA}}$; (b) broadening of $\Delta r_{\text{azi,eff}}$.

It is seen that, both $\Delta ISLR_{d\text{RMA}}$ and $F_{\Delta r_{\text{azi}}}^{\Delta v_x}$ increase linearly with $\frac{\Delta v_x}{v}$ beyond about $\frac{\Delta v_x}{v} = 3\,\%$. However, within the region of $0 \leqslant \frac{\Delta v_x}{v} < 3\,\%$ $\Delta ISLR_{d\text{RMA}}$ and $F_{\Delta r_{\text{azi}}}^{\Delta v_x}$ increase as a quadratic function of $\frac{\Delta v_x}{v}$, which is not in agreement with the property of QPE as indicated in (3.25). The reason of this discrepancy is the annular shape of the support band as depicted in Figure 2.8c. As illustrated in Figure 3.10, due to the annular support band, the azimuth broadening effect of IRF caused by QPE is not uniform but with one end bigger than the other end in the range direction. This uneven broadening causes the quadratic increase style of $\Delta ISLR_{d\text{RMA}}$ and $F_{\Delta r_{\text{azi}}}^{\Delta v_x}$ at low $\frac{\Delta v_x}{v}$ as illustrated in Figure 3.10a. As shown in Figure 3.10b, as $\frac{\Delta v_x}{v}$ increases, more energy leaks outside of the nominal mainlobe. Thus the uneven broadening effect within the mainlobe decreases and consequently results in the linear increase of $\Delta ISLR_{d\text{RMA}}$ and $F_{\Delta r_{\text{azi}}}^{\Delta v_x}$. According to Figure 3.9 for targets of $y_t = 1\,\text{m}$, $\frac{\Delta v_x}{v}$ should not be beyond 2.5 % to keep $F_{\Delta r_{\text{azi}}}^{\Delta v_x} < 1.2$.

According to (3.23), the influence of Δv_x is approximately proportional to $k_{rc} \cdot y_t \cdot \sin^2(2\psi)$. Therefore the requirement for $\frac{\Delta v_x}{v}$ can be deduced according to the presented results approximately. For a SAR system of different working range and carrier frequency, such as for $f_c = 77\,\text{GHz}$, the equivalent $\frac{\Delta v_x}{v}$ that has the same effect on the

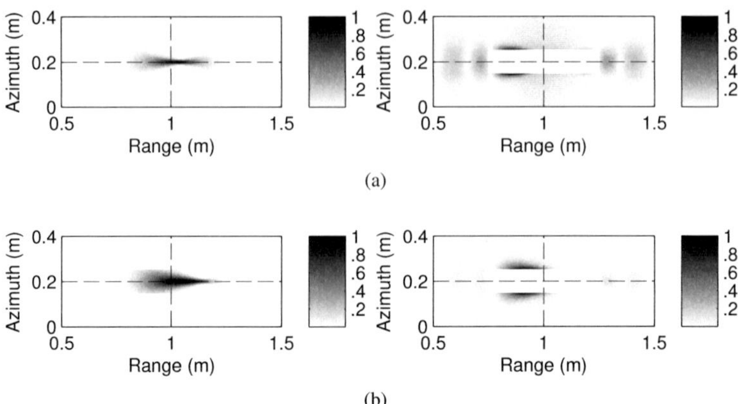

Figure 3.10: The mainlobes (left) and sidelobes (right) of IRF obtained when: (a) $\frac{\Delta v_x}{v} = 2\%$; (b) $\frac{\Delta v_x}{v} = 7\%$. (All figures are normalized with their own strongest points.)

SAR image can be calculated as

$$\left.\frac{\Delta v_x}{v}\right|_{f_c=77\,\mathrm{GHz}} = \frac{24 \cdot \sin^2(40°)}{77 \cdot \sin^2(20°)} \cdot \left.\frac{\Delta v_x}{v}\right|_{f_c=24\,\mathrm{GHz}} \approx 1.1 \cdot \left.\frac{\Delta v_x}{v}\right|_{f_c=24\,\mathrm{GHz}}, \tag{3.32}$$

and for $f_c = 120\,\mathrm{GHz}$, the equivalent value is

$$\left.\frac{\Delta v_x}{v}\right|_{f_c=120\,\mathrm{GHz}} = \frac{24 \cdot \sin^2(40°)}{120 \cdot \sin^2(20°)} \cdot \left.\frac{\Delta v_x}{v}\right|_{f_c=24\,\mathrm{GHz}} \approx 0.7 \cdot \left.\frac{\Delta v_x}{v}\right|_{f_c=24\,\mathrm{GHz}}. \tag{3.33}$$

It is interesting to see that Δv_x has the least influence on the SAR operating at $77\,\mathrm{GHz}$ among the three systems of different f_c with optimum Ψ, though the differences are not large.

3.3.3 Quadratic Motion Error

If a constant acceleration, Δa_x, exists, $\tilde{v}_x$ can be expressed as

$$\tilde{v}_x = v_0 + \Delta a_x t, \tag{3.34}$$

where v_0 is the initial velocity and the nominal $v = v_0 + \Delta a_x \cdot \frac{T_{L_A}}{2}$ is chosen as the middle value of $\tilde{v}_x$. By analogy with (3.22) and given $\tan(\psi) = \frac{vt}{y_t}$, the phase error related to azimuth compression can be obtained as

$$
\begin{aligned}
\Delta\phi_{\text{azicom}} &= \frac{k_r^2}{4k_{rc}} \frac{y_t}{v} \sin^2(2\psi) \cdot \Delta a_x t \\
&= \frac{y_t^2 \Delta a_x}{k_r k_{rc} v^2} \cdot k_x^3 \cdot \cos(\psi) \\
&\approx \frac{y_t^2 \Delta a_x}{k_r k_{rc} v^2} \cdot k_x^3,
\end{aligned}
\tag{3.35}
$$

which is a quasi-cubic function of k_x instead of quasi-quadratic function as (3.24). The effect of a cubic phase error on the SAR image causes asymmetric sidelobes of the system's IRF, as stated in [18]. However, the asymmetric sidelobes can also be considered as a broadening of the mainlobe in the sense of effective resolution as discussed in Subsection 2.4.3. Based on this, the effect of Δa_x or the requirement of it can be analyzed quantitatively. As presented in Appendix B, the expression of Δa_x which causes the same broadening of the mainlobe as Δv_x does can be written as

$$\Delta a_x = \frac{4}{3} \cdot \frac{v^2}{y_t \sin\left(\frac{\Psi}{2}\right)} \cdot \frac{\Delta v_x}{v}. \tag{3.36}$$

Given the requirement of $\frac{\Delta v_x}{v}$ in the sense of the broadening of $\Delta r_{\text{azi,eff}}$ according to Figure 3.9, the requirement of Δa_x can be obtained, which is relaxed proportionally to v^2.

The phase error caused by Δa_x deteriorates the SAR image along range direction via the term k_r in (3.35). For the SAR systems being considered in this dissertation, the change of (3.35) along k_r axis is trivial, e.g. for $f_c = 24\,\text{GHz}$ and $B = 1\,\text{GHz}$, $\frac{k_{r,\text{max}}}{k_{r,\text{min}}} \approx$ 1.04, which only introduces a phase error in the final SAR image. Hence this effect is not of interest in the imaging application and is not addressed here.

3.3.4 Sinusoidal Motion Error

The uneven sampling intervals along the azimuth direction caused by the sinusoidal vibrations depend on both their amplitude and frequency, the combination of which is infinite. Thus it is impractical to analytically investigate their influence. However, there is the international standard, ISO 2631, that gives a certain range of amplitude and frequency of vibrations with regard to human's comfort and perception. It is reasonable to choose motion parameters according to this standard when investigating the SAR system mounted on manned vehicles.

As described in Appendix A, 6 different degrees of discomfort (DoD) are chosen for the investigation, where the total values of the vibration, a_v, are $0\,\mathrm{m/s^2}$, $0.4725\,\mathrm{m/s^2}$, $0.75\,\mathrm{m/s^2}$, $1.2\,\mathrm{m/s^2}$, $1.875\,\mathrm{m/s^2}$ and $2\,\mathrm{m/s^2}$ respectively. The design idea of motion parameters is to consider the worst case situation. Intuitively, quadratic or higher orders vibrations cause worse degradation on the final result than Δv_x and Δa_x. Therefore a quadratic vibration is designed with respect to targets of $y_\mathrm{t} = 1\,\mathrm{m}$ with the vibration frequency as

$$f_x = 0.5 f_\mathrm{syn} = \left.\frac{v}{2L_\mathrm{t}}\right|_{y_\mathrm{t}=1\,\mathrm{m}} = \frac{v}{4\tan\left(\frac{\Psi}{2}\right)}. \tag{3.37}$$

Such motion errors with respect to targets of $y_\mathrm{t} < 1\,\mathrm{m}$ are equivalent to Δv_x or Δa_x, which have been analyzed above, while to targets of $y_\mathrm{t} > 1\,\mathrm{m}$ they are quadratic, cubic or even sinusoidal vibrations of multiple cycles, which result worse effects. According to Appendix A the amplitude of the acceleration is given as

$$A_{ax} = \frac{\sqrt{2}a_\mathrm{v}}{W_\mathrm{d}\left(f_x\right)}, \tag{3.38}$$

so the amplitude of the motion errors can be obtained as

$$A_x = \frac{A_{ax}}{\left(2\pi f_x\right)^2}. \tag{3.39}$$

Then, for $v = 1$, 2.5, 5, 10, 15 and $20\,\mathrm{m/s}$, the corresponding vibration motion errors are applied to generate raw SAR data in the simulator for targets with $y_\mathrm{t} = 1$, 2, 5, 10 and $15\,\mathrm{m}$ respectively. The effective $ISLR_{d\mathrm{RMA}}$ are determined with the mainlobe defined via $d_\mathrm{azi}^\mathrm{RMA}$ from motion-error-free data, and $\Delta r_\mathrm{azi,eff}$ are determined via $IRPR^\mathrm{RMA}$ which is also obtained from the motion-error-free data as described in subsection 2.4.3.

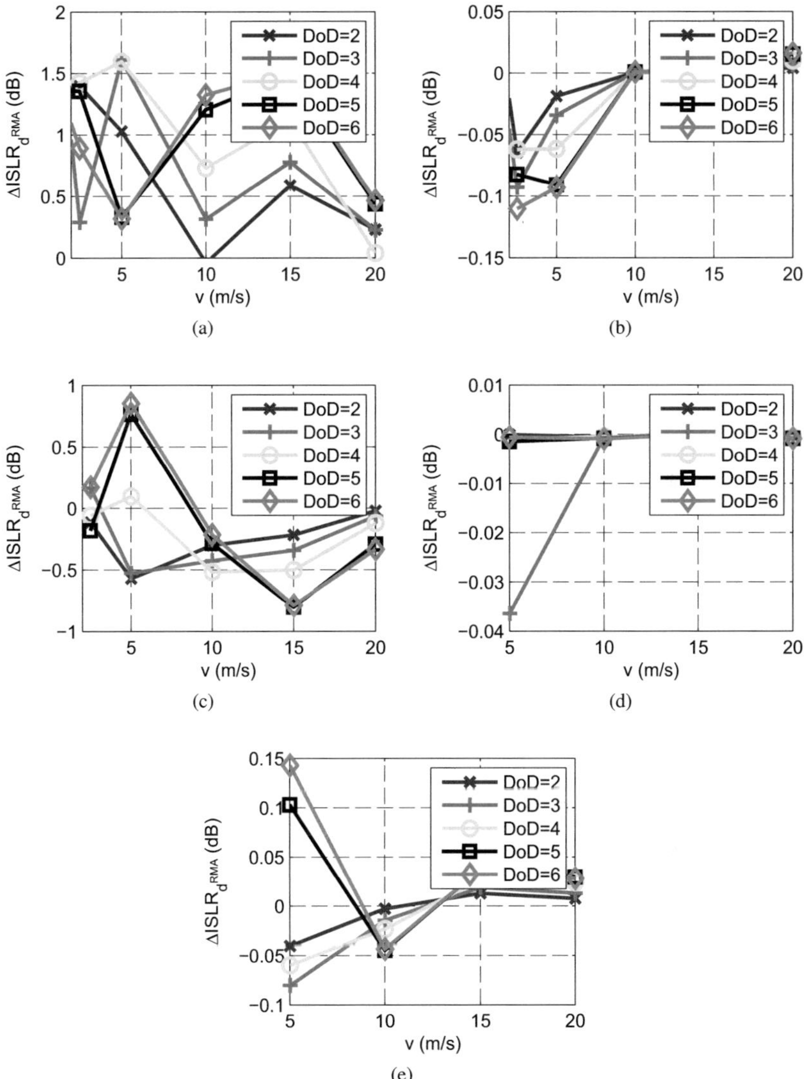

Figure 3.11: Changes of $ISLR_{d\text{RMA}}$ as a function of DoD and v: (a) $y_t = 1\,\text{m}$; (b) $y_t = 2\,\text{m}$; (c) $y_t = 5\,\text{m}$; (d) $y_t = 10\,\text{m}$; (e) $y_t = 15\,\text{m}$.

The differences between the motion-error-deteriorated and motion-error-free $ISLR_{d\text{RMA}}$ are depicted in Figure 3.11. On the one hand, for a target at $y_t = 1\,\text{m}$, the applied vibration is quadratic in the target's synthetic aperture, whose effect with regard to the degradation of $ISLR_{d\text{RMA}}$ is relatively obvious, but still less than $2\,\text{dB}$ worse than the motion-error-free $ISLR_{d\text{RMA}}$. On the other hand, for targets at $y_t > 1\,\text{m}$, the applied vibration is a sinusoid with at least one cycle in the synthetic aperture. Its effect on the deterioration of $ISLR_{d\text{RMA}}$ is negligible or even decreases the $ISLR_{d\text{RMA}}$ in some cases. In general, the leaking energy caused by uneven sampling spreads mainly within the effective mainlobe, so the resulting degradation of $ISLR_{d\text{RMA}}$ can be ignored.

The effects of the sinusoidal vibration along the azimuth direction as deteriorating $\Delta r_{\text{azi,eff}}$ are depicted in Figure 3.12. It can be seen that the broadening factor of $\Delta r_{\text{azi,eff}}$ is proportional to y_t and inversely proportional to v. Based on these results, the minimum velocity required when no motion compensation is performed can be deduced. For example, when no motion compensation along the azimuth direction is applied, in order to keep the broadening factor below 1.2 for DoD of 4 (the point where the passenger starts to feel dis-comfortable), the vehicle's velocity should be higher than $10\,\text{m/s}$.

Since the influence of motion error is proportional to f_c, as indicated by (3.23), (3.27) and (3.29), a higher minimum v is required for SAR with $f_c = 77\,\text{GHz}$ or $f_c = 120\,\text{GHz}$, which can be determined with the same procedure as proposed above.

3.4 Range

For the ease of explanation, targets of interest are assumed at the same height level with the radar. Therefore targets are in the same plane with range motion errors, $\Delta y(t)$, and the coordinates of the radar and targets are expressed in the 2D plane, i.e. IDP. However, the analysis performed and results obtained in this section also apply to motion errors in the slant range plane which is formed by the nominal slant range vectors with regard to a certain target.

As discussed in Appendix C, the SAR system is an approximate linear system with respect to the motion errors, Δr_{in}, so the influences of $\Delta y(t)$ are analyzed with respect to each component in (3.10) in this section respectively. As illustrated in

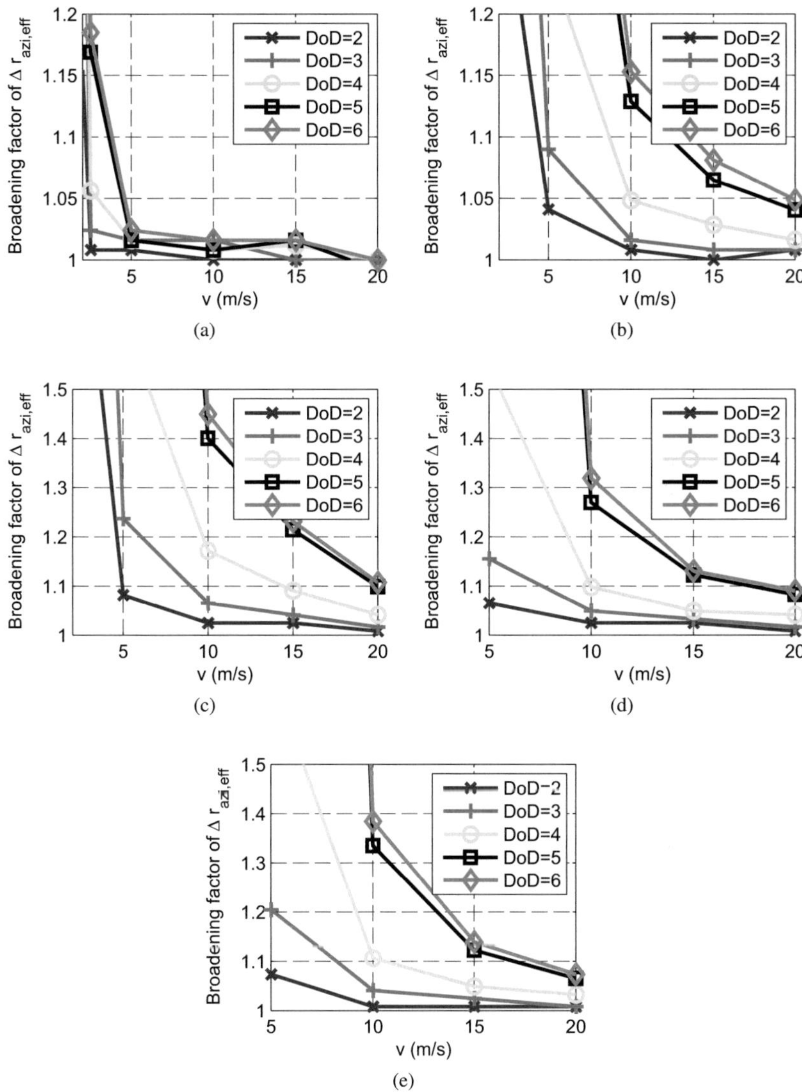

Figure 3.12: The broadening factor of $\Delta r_{\mathrm{azi,eff}}^{\mathrm{RMA}}$ as a function of DoD and v: (a) $y_t = 1\,\mathrm{m}$; (b) $y_t = 2\,\mathrm{m}$; (c) $y_t = 5\,\mathrm{m}$; (d) $y_t = 10\,\mathrm{m}$; (e) $y_t = 15\,\mathrm{m}$.

subsection 3.1.3, the linear artificial motion errors can be decomposed into a constant deviation in parallel with the nominal trajectory and a linear deviation trajectory, which correspond to terms Δy_0 and $\Delta v_y t$ in (3.10) respectively. Their influences with and without first-order motion compensation are compared in subsection 3.4.1 and 3.4.2 respectively. Furthermore, the results with first-order motion compensation are equivalent to the influences of residual motion errors which are caused by non-ideal motion measurements or estimations. The influences of $\frac{1}{2}\Delta a_y \cdot t^2$ and $\sum_i A_{yi} \cdot \sin(2\pi f_{yi} t)$ are investigated by the SAR simulator, and evaluated via Δr_{azi} and *ISLR*. All the parameters investigated are designed according to the DoD as described in Appendix A, by analogy with subsection 3.3.4.

3.4.1 Constant Position Error

The constant position error Δy_0 is a kind of artificial motion error as defined in subsection 3.1.3. It is part of the real motion errors for far range targets but unnecessary to be compensated for near range targets. Therefore the following analysis focuses on the influences resulting from performing first-order motion compensation for Δy_0 with regard to the near range targets.

The Δy_0 deteriorated IF data can be expressed as

$$\tilde{s}_{\mathrm{IF}}(x_{\mathrm{am}}, k_r) = A_t w_r(k_r) w_a(\tilde{r}_t) \exp\left[-jk_r \cdot \sqrt{(x_{\mathrm{am}} + v\hat{t} - x_t)^2 + (\Delta y_0 - y_t)^2}\right]. \quad (3.40)$$

In (3.40), the errors in $w_a(\tilde{r}_t)$ are ignored since Δy_0 only causes a smooth change in the amplitude weighting function, $w_a(r_t)$, whose influence in the spatial frequency domain after azimuth Fourier transformation is trivial as opposed to the exponential term, $\exp[\cdot]$. Without motion compensation, the only effect of Δy_0 is to change the coordinates of the target in the SAR image from (x_t, y_t) to $(x_t, y_t - \Delta y_0)$, which, however, actually reflect the real relative distance between the radar and the target.

In the following discussion, the real range of the closest approach is expressed as $y'_t = y_t - \Delta y_0$, where the superscript "'" denotes the term is represented in the $x'y'$-coordinate system as illustrated in Figure 3.6. So the same IF data expressed in the

$x'y'$-coordinate system is

$$s'_{\text{IF}}(x_{\text{am}}, k_r) = A_t w_r(k_r) w_a(r'_t) \exp\left[-jk_r \cdot \sqrt{(x_{\text{am}} + v\hat{t} - x_t)^2 + (y'_t)^2}\right]. \qquad (3.41)$$

If motion compensation is performed, after compensating the space-invariant motion errors (the first-order motion compensation) via (3.7), the IF data can be expressed as

$$\begin{aligned}
\tilde{s}_{\text{IF,in}}(x_{\text{am}}, k_r) &= \tilde{s}_{\text{IF}}(x_{\text{am}}, k_r) \cdot \exp(jk_r \cdot \Delta r_{\text{in}}) \\
&= s_{\text{IF}}(x_{\text{am}}, k_r) \cdot \exp[-jk_r \cdot \Delta r_{\text{var,t}}(\Delta y_0, x_t, y_t)].
\end{aligned} \qquad (3.42)$$

Here, $\Delta r_{\text{var,t}}(\Delta y_0, x_t, y_t)$ is the space-variant component containing the motion errors corresponding to Δy_0 and with respect to a certain target located at (x_t, y_t). If further space-variant motion compensation is performed to eliminate the term $\exp[-jk_r \cdot \Delta r_{\text{var}}(\Delta y_0, x_t, y_t)]$ in (3.42), the target will be registered to the nominal position (x_t, y_t) in the final SAR image. However, as mentioned in subsection 3.1.3 such two-order motion compensation algorithms are usually computationally expensive and also not perfect due to the contradiction between the SAR processing in frequency domain and compensating motion errors changing in space domain. Therefore, the knowledge of allowable levels of Δy_0 is required to determine whether further motion compensation is needed.

By expanding $\tilde{r}_t$ and r_t at the second-order, $\Delta r_{\text{var}}(\Delta y_0, x_t, y_t)$ can be calculated as

$$\begin{aligned}
\Delta r_{\text{var}}(\Delta y_0, x_t, y_t) &= \tilde{r}_t - r_t - \Delta r_{\text{in}} \\
&\approx y_t + \Delta y_0 + \frac{(x_{\text{am}} + v\hat{t} - x_t)^2}{2(y_t - \Delta y_0)} - y_t - \frac{(x_{\text{am}} + v\hat{t} - x_t)^2}{2y_t} + \Delta y_0 \\
&= \frac{\Delta y_0}{2(y_t - \Delta y_0)y_t} \cdot (x_{\text{am}} + v\hat{t} - x_t)^2,
\end{aligned} \qquad (3.43)$$

which is a quadratic function of x. Substituting (3.43) into (3.42) and performing an azimuth FFT on it, the IF data are then transformed to the 2D spatial frequency. Since the fluctuation of $\Delta r_{\text{var,t}}(\Delta y_0, x_t, y_t)$ is much smaller than the fluctuation of the phase in $s_{\text{IF}}(x_{\text{am}}, k_r)$, according to POSP the term $\exp[-jk_r \cdot \Delta r_{\text{var}}(\Delta y_0, x_t, y_t)]$ is transformed into the spatial frequency domain just by substituting a variable using (2.47), leading to

$$\exp\left[-jk_r \cdot \Delta r_{\text{var}}\left(\Delta y_0\right)\right] \;=\; \exp\left[-jk_r \cdot \frac{\Delta y_0}{2\left(y_t - \Delta y_0\right)y_t}\cdot\left(\frac{y_t k_x}{\sqrt{k_r^2 - k_x^2}}\right)^2\right]$$

$$=\; \exp\left[-j\cdot\frac{\Delta y_0 y_t}{2\left(y_t - \Delta y_0\right)}\cdot\frac{\sqrt{k_x^2 + k_y^2}}{k_y^2}\cdot k_x^2\right]. \qquad (3.44)$$

Note that (3.44) is not a QPE in k_x domain though it results from a QPE in x domain. Instead, (3.44) is a function of both k_x and k_y , which deteriorates the IRF in both azimuth and range directions. In addition, the targets of certain *real* range of the closest approach, y_t', suffer from the residual motion errors. Therefore the phase error term should be expressed as a function of y_t'. Subsequently (3.44) can be expressed as

$$\exp\left[-jk_r \cdot \Delta r_{\text{var}}\left(\Delta y_0\right)\right] = \exp\left[-j\cdot\frac{\Delta y_0\left(y_t' + \Delta y_0\right)}{2y_t'}\cdot\frac{\sqrt{k_x^2 + k_y^2}}{k_y^2}\cdot k_x^2\right]. \qquad (3.45)$$

It can seen that the influence of $\Delta r_{\text{var}}\left(\Delta y_0, x_t, y_t'\right)$ is independent of y_t', and linearly proportional to Δy_0 when $y_t' \gg \Delta y_0$. In addition, since $\sqrt{k_x^2 + k_y^2} = k_r = \frac{4\pi}{c}\left(f_{\min} + \gamma\hat{t}\right)$, the influence of Δy_0 is also proportional to the carrier frequency of the SAR system.

Since it is difficult to obtain a closed expression of the influence caused by (3.45) for any y_t', by analogy with the analysis in section 3.3, computational simulations according to the DoD are used to perform the investigation.

First of all, the range of y_t' should be determined. The range swath of SAR system in near-range applications is

$$\frac{2D^2}{\lambda_{\min}} < r_{t,0} < r_{\max}. \qquad (3.46)$$

In (3.46), $\frac{2D^2}{\lambda_{\min}}$ is the far-field boundary of the antenna as defined in [122], where D ($D > \lambda_{\max}$) is the maximum overall dimension of the antenna and $\lambda_{\min}$ is the minimum wavelength of the transmitted chirp; $r_{\max}$ is the maximum working range of the SAR system as defined in subsection 2.3.4. Given that $D = 4.2\,\text{cm}$ and $\lambda_{\min} = 1.28\,\text{cm}$ for the antennas used in our radar demonstrator, the nearest range of the working range swath of the SAR, $\frac{2D^2}{\lambda_{\min}}$, is about 28 cm. Here $y_{t,\text{Nearest}} = 0.5\,\text{m}$ is chosen as the nearest range of interest and $y_{t,\text{Furthest}} = r_{\max} = 16\,\text{m}$ as the furthest range of interest.

Then the range of Δy_0 should be decided. The possible absolute values of Δy_0 are actually the amplitudes of the high-order motion errors with regard to the target of $y_{t,\text{Furthest}}$. To obtain a reasonable range of Δy_0 for simulation, the following steps are performed.

1. Since the corresponding MSR_{Nearest} of the motion errors with respect to the nearest targets should be smaller than 0.25 as defined in (3.6), accordingly the upper limit of the corresponding MSR_{Furthest} of the same motion errors with respect to the furthest targets can be obtained as

$$MSR_{\text{Furthest}} = \frac{L_{t,\text{Furthest}}}{L_{t,\text{Nearest}}} \cdot MSR_{\text{Nearest}} < \frac{16}{0.5} \cdot \frac{1}{4} = 8. \tag{3.47}$$

The lower limit of MSR_{Furthest} is 0.5, which makes sure that the motion errors investigated here are quadratic or higher-order vibrations that need to be compensated for the furthest targets. A series of MSR_{Furthest} are given in the range of $MSR_{\text{Furthest}} \in [0.5, 8]$ with steps of 0.1.

2. A series of possible v are given as $v = 1, 2, 3, ..., 24, 25\,\text{m/s}$.

3. According to (3.6), for each v, a series of f_y are calculated using the given series of MSR_{Furthest} as

$$f_y = \frac{v}{L_{t,\text{Furthest}}} \cdot MSR_{\text{Furthest}}, \tag{3.48}$$

and the corresponding values of $W_d(f_y)$ can be determined as described in Appendix A.

4. By analogy with the analysis in subsection 3.3.4, 5 different DoD are given with a_v being $0.4725\,\text{m/s}^2$, $0.75\,\text{m/s}^2$, $1.2\,\text{m/s}^2$, $1.875\,\text{m/s}^2$ and $2\,\text{m/s}^2$ for DoD of 2, 3, 4 and 5 respectively

5. Substituting all the combinations v, W_d and a_v into (A.7), the range of the A_a, subsequently the range of the absolute values of Δy_0 can be obtained, where Δy_0 is calculated as

$$|\Delta y_0| = \frac{A_a}{(2\pi f_y)^2}. \tag{3.49}$$

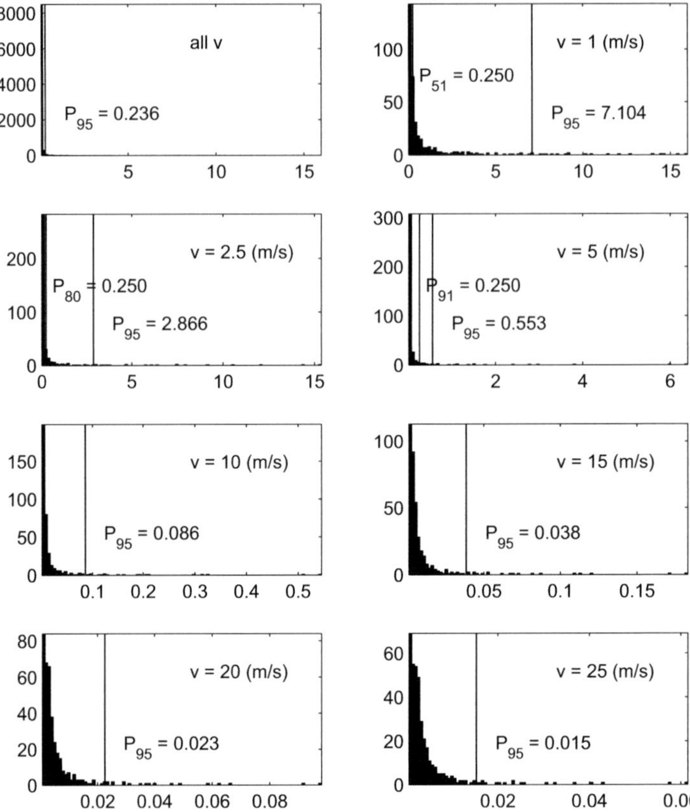

Figure 3.13: The histograms of Δy_0 under different circumstances in terms of v and DoD, where the horizontal axis is the value of $|\Delta y_0|$ in unit of m and the vertical axis is the number of samples.

The histograms of the obtained $|\Delta y_0|$ are shown in Figure 3.13 for different v respectively, where the horizontal axis is the value of $|\Delta y_0|$ in unit of m and the vertical axis is the number of samples. In each histogram, all DoD at the same v are included, and the value of the 95th percentile, P_{95}, is labeled, which means 95% of $|\Delta y_0|$ fall below P_{95}. Some comments are now in order.

First, some f_y are too low to be covered by ISO 2631 as depicted in Figure A.2, $|\Delta y_0|$ for such f_y are not included in the histograms in Figure 3.13. In addition, it is intuitive that $|\Delta y_0|$ that are larger than $y_{t,\text{Furthest}}$ should not be used either. The reason of the presence of such large $|\Delta y_0|$ is as follows. ISO 2631 is originally proposed for evaluation with regard to the human's perception of vibration. Since a human is much less sensitive for low frequency vibration, much larger amplitude of the low frequency vibration is allowed for a certain DoD than that of other frequency vibrations. Nevertheless, these two parts of $|\Delta y_0|$ are only about 0.8% of the total $|\Delta y_0|$, so most of the $|\Delta y_0|$ obtained by the proposed method are reasonable.

Second, it is interesting to see that the value of $|\Delta y_0|$ that can be considered as artificial motion errors is inversely proportional to v. Though the phase errors caused by $|\Delta y_0|$ is independent of v as indicated in (3.45), a higher v will restrict the possible maximum $|\Delta y_0|$ at a certain DoD. Therefore near range targets are less likely to suffer from artificial motion errors when the SAR moves at higher velocity.

Third, since 95% of the whole data set of $|\Delta y_0|$ are smaller than $0.24\,\text{m}$ as indicated in the first row of the first column of Figure 3.13, it is adequate to set the range of Δy_0 for simulations as $\Delta y_0 \in [-0.25, 0.25]\,\text{m}$. The corresponding percentage of $|\Delta y_0| < 0.25\,\text{m}$ for $v = 1, 2.5$ and $5\,\text{m/s}$ are 51%, 80% and 91% respectively, as labeled in the corresponding histograms.

Given the range of Δy_0, a series of simulations have been performed for the nearest target of $y_{t'} = 0.5\,\text{m}$, which is kept constant for all the simulations. The results are shown in Figure 3.14. The differences between the motion-error-deteriorated and motion-error-free $ISLR_{d\text{RMA}}$ are depicted in Figure 3.14a. It is seen that $\Delta ISLR_{d\text{RMA}}$ is symmetric about $\Delta y_0 = 0$, and smaller than $1\,\text{dB}$ below about $|\Delta y_0| = 15\,\text{cm}$. Within this region, the deterioration happens mainly within the mainlobe defined via $d_{\text{azi}}^{\text{RMA}}$ from motion-error-free data. While beyond $|\Delta y_0| = 15\,\text{cm}$, $\Delta ISLR_{d\text{RMA}}$ increases linearly to $|\Delta y_0|$ at great speed. The target is expected to be registered to y_t in IDP after performing motion compensation, the difference between the range coordinate of the strongest point in the SAR image, $\tilde{y}_t$, and the nominal y_t is determined and denoted with $\Delta y_t = \tilde{y}_t - y_t$. As also shown in Figure 3.14a, Δy_t is odd symmetric about $\Delta y_0 = 0$ and keeps a roughly constant slope within the region of $|\Delta y_0| < 15\,\text{cm}$. As indicated in (3.45) the influence of $\Delta r_{\text{var}}(\Delta y_0, x_t, y_t')$ is independent of $y_{t'}$ within the region of $|\Delta y_0| \ll y_t'$. Theoretically this linear relationship between Δy_t and Δy_0 can be used for registering the targets to their nominal positions even without second-order motion compensation.

In practical cases, however, there are two problems which make this intuitive position correcting method useless. First, all the equations derived in this subsection are based on the assumption that Δy_0 exists in the whole synthetic aperture of a certain target. In order to perform the position correction, the targets that meet this assumption should be determined first, which can be complex, since Δy_0 is a variable that depends on the target's coordinates, (x_t, y_t). Second, as shown in Figure 3.14b, the broadening factor of $\Delta r_{azi,eff}$, $F_{\Delta r_{azi}}^{\Delta y_0}$, grows much faster than $\Delta ISLR_{dRMA}$ and Δy_t, in the region where $F_{\Delta r_{azi}}^{\Delta y_0}$ is smaller than 1.5, $|\Delta y_t|$ is smaller than 5 cm, which is only about 33% of Δr_{ran} given a radar with $B = 1\,\mathrm{GHz}$. So this position correction method is not worth the computational cost compared to the deterioration it would correct.

Intrinsically, applying first-order motion compensation on $\tilde{s}_{IF}(x_{am}, k_r)$ as performed by (3.42) is equivalent to introducing a phase error term caused by residual motion errors i.e. the motion measurement errors, of the same value of Δy_0 on the motion-error-free SAR data in $x'y'$-coordinate system, $s'_{IF}(x_{am}, k_r)$, which can be expressed as

$$\tilde{s}'_{IF}(x_{am}, k_r) = s'_{IF}(x_{am}, k_r) \cdot \exp\left(-jk_r \cdot \Delta y_0\right). \tag{3.50}$$

So as discussed above, the residual motion errors will shift the target along range direction with a value of $\Delta y_0 + \Delta y_t$ in $x'y'$-coordinate system, and broaden the azimuth resolution in proportional to $|\Delta y_0|$.

It should be noted that for the cases of $y'_t > 0.5\,\mathrm{m}$, the influences illustrated in Figure 3.14 may change in the region where $|\Delta y_0|$ is comparable to $y'_t = 0.5\,\mathrm{m}$. However, within the region of $|\Delta y_0| < 12\,\mathrm{cm}$, the influences remain almost unchanged according to the simulations, which are not shown here in the interest of brevity. Such independence with respect to y'_t is also dictated mathematically by (3.45). Therefore the relationship between $F_{\Delta r_{azi}}^{\Delta y_0}$ and $|\Delta y_0|$ listed in Table 3.1 in the case of $y'_t = 0.5\,\mathrm{m}$ can be used as common allowable residual motion errors levels for any y'_t when designing motion compensation algorithms.

3.4.2 Constant Velocity Error

In this subsection, the constant range velocity component, Δv_y, in (3.10) is considered. By analogy with the analysis of Δy_0, the influence of Δv_y is investigated by comparing the results with and without first-order motion compensation. As shown in

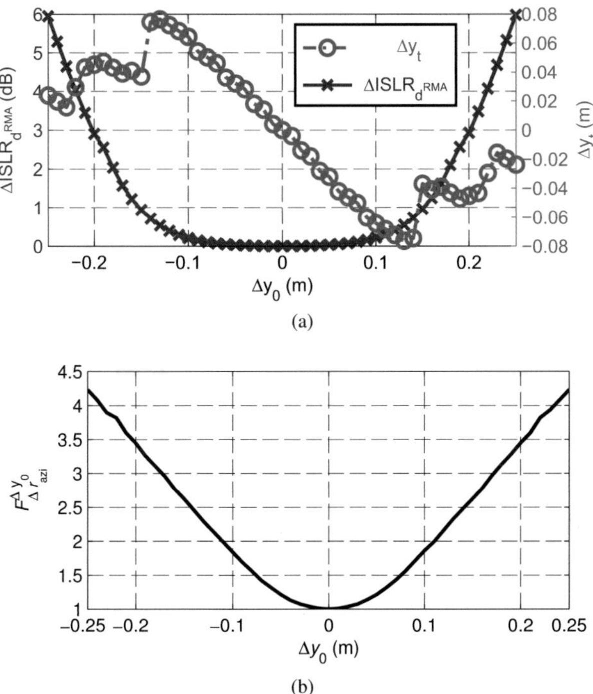

Figure 3.14: The influences of Δy_0 in the case of $f_c = 24\,\text{GHz}$, $\Psi = 40°$ and $y_{t'} = 0.5\,\text{m}$: (a) increasing $ISLR_{d^\text{RMA}}$ and not perfect shift in range direction; (b) broadening $\Delta r_{\text{azi,eff}}$.

Table 3.1: Effects of Δy_0 on $\Delta r_{\text{azi,eff}}$

$F_{\Delta r_{\text{azi}}}^{\Delta y_0}$	$|\Delta y_0|$ (cm)
1.1	3.4
1.2	4.8
1.4	6.7
1.6	8.3
1.8	9.6
2.0	11.1

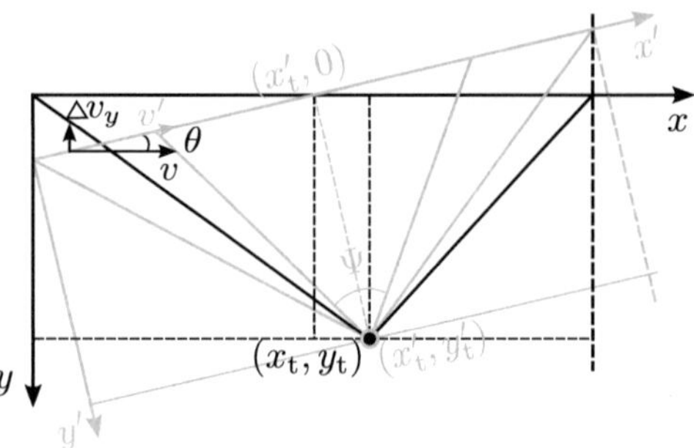

Figure 3.15: Geometry of constant range velocity error, $\Delta v_y < 0$, where the nominal xy-coordinate system is in black and the $x'y'$-coordinate system defined with regard to the short-term trajectory is in gray.

Figure 3.15, the geometry of SAR data collecting in the presence of a constant linear velocity component in the nominal range direction is illustrated in detail, where the quantities related to the nominal xy-coordinate system are in black while the quantities related to the $x'y'$-coordinate system defined with regard to the short-term trajectory are in gray. The effective velocity of the radar, v', in $x'y'$-coordinate system is assumed to be constant, and Δv_y is the artificial error that is introduced by defining the nominal trajectory at an angle of θ to the short-term trajectory of a certain target located at (x'_t, y'_t) in $x'y'$-coordinate system. Accordingly, the relationships between the nominal velocity, v, Δv_y and v' can be expressed as

$$v = v' \cdot \cos \theta, \tag{3.51}$$

$$\Delta v_y = v' \cdot \sin \theta. \tag{3.52}$$

In the case illustrated in Figure 3.15, Δv_y is negative and so is θ.

It is well known that a linear phase error in the spectrum of SAR data will cause the target to shift in azimuth direction. In the literature, e.g. [18, 123], the link between the linear motion error in space domain to the linear phase error in spatial frequency domain is presumed. However, this assumption only exists for a SAR system with narrow Ψ and with additional approximations. The effects of Δv_y have been analyzed in [60] in a more complete way, where a constant range velocity error introduced by performing first-order motion compensation is transformed to the spatial frequency domain. Note that though such motion error is called "residual error" in [60], it is essentially the artificial linear motion error discussed here. The analysis showed that besides a shift along azimuth direction, a shift in range direction is caused by Δv_y as well. However, the analysis also presumed the usage of narrow Ψ, which thus cannot cover the cases discussed in this dissertation. Furthermore, the author overlooked the difference between v' and v. As will be revealed later, such ignorance will cause further deterioration in the SAR image. At variance to the analysis in [60], an intuitive way to analyze the influences of Δv_y is presented in the following paragraphs, which takes into account the above mentioned deficiencies.

The IF data only containing a $\Delta v_y t$ motion error component can be written as

$$\tilde{s}_{\text{IF}}(x_{\text{am}}, k_r) = A_t w_r(k_r) w_a(\tilde{r}_t) \exp\left[-jk_r \cdot \sqrt{(x_{\text{am}} + v\hat{t} - x_t)^2 + (\Delta v_y t - y_t)^2}\right]. \quad (3.53)$$

First of all, the effect of aliasing is considered. As depicted in Figure 3.15, in the nominal coordinate system, the squint angle of the SAR system is no longer zero but becomes θ. In this case, the original *PRF* that is designed according to Ψ and the real velocity v' as dictated in (2.21) may not meet the Nyquist criterion in azimuth direction, subsequently aliasing along azimuth may appear. As illustrated in Figure 3.16a, the spectrum of the SAR data is shifted along k_x axis. Since the squint angle $\theta = 18°$ is much larger than $\frac{\Psi}{2} = 10°$, aliasing happens in the azimuth spectral domain. As shown in Figure 3.16b, even after performing motion compensation with regard to Δr_t, a weaker point still appears as a result of the spectral aliasing. Since such deterioration is caused intrinsically by violating the Nyquist criterion, it cannot be compensated by the second-order motion compensation.

In addition, as the support band is annular when RMA is used, the spectrum shifted by θ falls into the side part of the annulus. Since for the same Doppler bandwidth the area of the side part is smaller than the middle part as illustrated in Figure 2.8c, the *SNR* of the SAR image is decreased. Furthermore, when an azimuth frequency window is applied, the shift of the spectrum will cause the useful information to be weakened by the weighting function, thereby decreasing the *SNR* of the SAR image further. The motion compensation in remote sensing SAR application updates the reference function in spatial frequency domain every second to take into account the variation of Doppler centroid [73], which is essentially moving the spatial frequency window in quasi-real-time. However, as discussed in subsection 3.1.3, the change of squint angle caused by

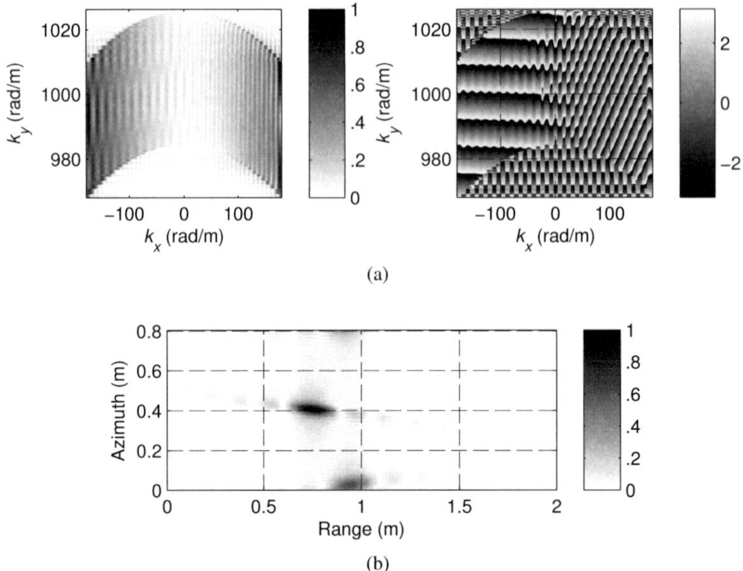

Figure 3.16: Aliasing caused by Δv_y in the case of $f_c = 24\,\text{GHz}$, $\Psi = 20°$ and $\theta = 18°$: (a) the amplitude and phase of the spatial support band; (b) the SAR image generated with motion compensation in regard to Δr_t.

Δv_y is specific for certain near range targets and only keeps constant within a small part of the whole SAR processing block. Therefore, such motion compensation performed in spatial frequency domain is not applicable for dealing with artificial motion errors in near-range applications.

Instead of performing analytical formula deduction as in [60], the influence of Δv_y via the exponential term in (3.53) can be investigated in the view of rotation of coordinate axes. Without motion compensation for the linear motion along nominal range direction in the xy-coordinate system, the SAR processing is essentially performed in the $x'y'$-coordinate system with the real constant velocity v' and without any motion error. The coordinates of the target in the $x'y'$-coordinate system denoted with (x'_t, y'_t) are obtained by rotating the the xy-coordinate system with an angle of θ, where the rotation angle is positive when the rotation is counterclockwise in a right-handed

68

Cartesian coordinate system. With the knowledge of linear algebra the coordinates of the same target in the nominal xy-coordinate system, (x_t, y_t), can be obtained via rotation matrix of coordinate axes and expressed as

$$x_t = x_t' \cdot \cos\theta - y_t' \cdot \sin\theta, \tag{3.54}$$

$$y_t = y_t' \cdot \cos\theta. \tag{3.55}$$

However, if only a first-order motion compensation is applied, the space-variant motion errors remaining in $\tilde{s}_{\text{IF,in}}(x_{\text{am}}, k_r)$ as indicated in (3.42) will deteriorate the SAR image analogously to the case of Δy_0. The space-variant motion errors can be calculated by substituting Δy_0 with $\Delta v_y t$ into (3.44) and expressed as

$$\Delta r_{\text{var}}(\Delta v_y, x_t, y_t) = \frac{\Delta v_y t}{2(y_t - \Delta v_y t) y_t} \cdot (x_{\text{am}} + v\hat{t} - x_t)^2. \tag{3.56}$$

By analogy with the case of Δy_0, according to POSP, $\exp\left[-jk_r \cdot \Delta r_{\text{var}}(\Delta v_y, x_t, y_t)\right]$ is transformed into the spatial frequency domain by a variable substituting using (2.47), and can be expressed as

$$\begin{aligned}
\exp\left[-jk_r \cdot \Delta r_{\text{var}}(\Delta v_y)\right] &= \exp\left[j \cdot k_r \cdot \frac{y_t \tan\theta \tan^3 \psi}{2(1 + \tan\theta \tan\psi)}\right] \\
&= \exp\left[j \cdot k_r \cdot \frac{y_t' \sin\theta \tan^3 \psi}{2(1 + \tan\theta \tan\psi)}\right].
\end{aligned} \tag{3.57}$$

In (3.57), the term $\tan\psi = \frac{k_x}{k_y}$ indicates that the deterioration is in both azimuth and range directions in the SAR image. However, as having been analyzed in subsection 3.3.3, the phase error related to k_y is quasi-constant and thus can be ignored for imaging applications. The term $\tan^3 \psi$ indicates that (3.57) is approximately a cubic phase error in k_x domain; the term $k_r = \frac{4\pi}{c}(f_{\text{min}} + \gamma\hat{t})$ indicates that the influence of Δv_y is proportional to the carrier frequency of the SAR system; the terms $\sin\theta = \frac{\Delta v_y}{v'}$ and $\tan\theta = \frac{\Delta v_y}{v}$ indicate that it is the ratio of velocity that determines the level of deterioration instead of the absolute Δv_y. It is seen that the deterioration is linear proportional to y_t' as opposed to the case of Δy_0 which is independent of y_t'. It should be noted that in [60] such space-variant motion errors are not taken into account since a narrow Ψ is presumed, which makes the phase in (3.57) approximating to zero. It is impractical to derive a closed expression of the effects resulting from (3.57). By analogy with the analysis of Δy_0, computation simulations for a series of possible θ with regard to a series of y_t' are performed.

The possible $\theta = \arctan\left(\frac{\Delta v_y}{v}\right)$ is the slope of a sinusoid which needs to be compensated for far range targets while part of it being the whole synthetic aperture for near range targets is almost linear. So the range of θ can be deduced by analogy with the deduction of the range of Δy_0. For reason of completion, a full procedure of the deduction of the range of θ is described as follows.

1. Since the corresponding MSR_{Nearest} of the motion errors with respect to the nearest targets should be smaller than 0.25 as defined in (3.6), accordingly the upper limit of the corresponding MSR_{Furthest} of the same motion errors with respect to the furthest targets can be obtained as

$$MSR_{\text{Furthest}} = \frac{L_{t,\text{Furthest}}}{L_{t,\text{Nearest}}} \cdot MSR_{\text{Nearest}} < \frac{16}{0.5} \cdot \frac{1}{4} = 8. \tag{3.58}$$

The lower limit of MSR_{Furthest} is 0.5, which makes sure that the motion errors investigated here are quadratic or higher-order vibrations that need to be compensated for the furthest targets. A series of MSR_{Furthest} within the obtained range, $MSR_{\text{Furthest}} \in [0.5, 8]$, are given with values of uniform interval of 0.1.

2. A series of v are given for investigation, i.e. $v = 1, 2, 3, ..., 24, 25\,\text{m/s}$.

3. According to (3.6), for each v, a series of f_y are calculated using the given series of MSR_{Furthest} as

$$f_y = \frac{v}{L_{t,\text{Furthest}}} \cdot MSR_{\text{Furthest}}, \tag{3.59}$$

and the corresponding values of $W_d(f_y)$ can be determined as described in Appendix A.

4. By analogy with the analysis in subsection 3.3.4, 5 different DoD are given with a_v being $0.4725\,\text{m/s}^2$, $0.75\,\text{m/s}^2$, $1.2\,\text{m/s}^2$, $1.875\,\text{m/s}^2$ and $2\,\text{m/s}^2$ for DoD of 2, 3, 4 and 5 respectively.

5. Substituting all the combinations v, W_d and a_v into (A.7), the range of the amplitude of the acceleration, A_a, subsequently the range of the amplitude of the

velocity, $|\Delta v_y|$, which is calculated as

$$|\Delta v_y| = \frac{A_a}{2\pi f_y},$$ (3.60)

can be obtained.

6. Finally, the range of θ for each nominal v can be obtained by

$$\theta = \arctan\left(\frac{\Delta v_y}{v}\right).$$ (3.61)

The histograms of the obtained $|\theta|$ are shown in Figure 3.17 for different v respectively, where the horizontal axis is the value of $|\theta|$ in unit of $^\circ$ and the vertical axis is the number of samples. In each histogram, all DoD at the same v are included, and the value of the 95th percentile, P_{95}, is labeled, which means 95% of $|\theta|$ fall below P_{95}. Some comments are now in order.

First, for the same reasons as stated in the analysis of Δy_0, some $|\theta|$ are not included in the histograms. The omitted $|\theta|$ are only about 1.4% of the total $|\theta|$, so most of the $|\theta|$ obtained by the proposed method are reasonable.

Second, though the phase error in spatial frequency domain is a function of $\tan\theta$ as dictated in (3.57), a higher nominal velocity, v, will restrict the possible maximum $|\theta|$, subsequently restrict the effects of Δv_y.

Third, since 95% of the whole data set of $|\theta|$ are smaller than 18.54° as indicated in the first row of the first column of Figure 3.13, it is adequate to set the range of $|\theta|$ for simulations as $\theta \in [-20, 20]^\circ$. The corresponding percentage of $|\theta| < 20^\circ$ for $v = 1$ and $2.5\,\mathrm{m/s}$ are 31% and 83% respectively, as labeled in the corresponding histograms.

Since the influences of Δv_y depend on y'_t, a series of simulations of different y'_t with the θ of the designated range have been performed, and the results are shown in Figure 3.18, 3.19 and 3.20.

It is seen in Figure 3.18a that the increase of $ISLR_{d\mathrm{RMA}}$ is proportional to both y'_t and θ, which is in agreement with (3.57). As shown in Figure 3.18b the broadening factor of $\Delta r_{\mathrm{azi,eff}}$, $F_{\Delta r_{\mathrm{azi}}}^{\Delta v_y}$, is proportional to both y'_t and θ too. The sawtooth-style of the lines

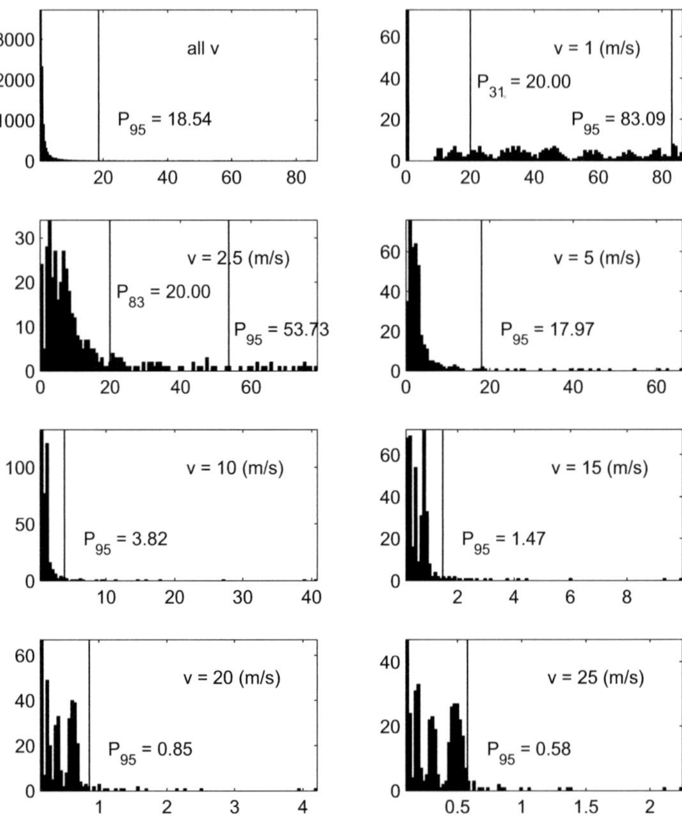

Figure 3.17: The histograms of $\theta = \arctan\left(\frac{\Delta v_y}{v}\right)$ under different circumstances in terms of v and DoD, where the horizontal axis is the value of $|\theta|$ in unit of $^\circ$ and the vertical axis is the number of samples.

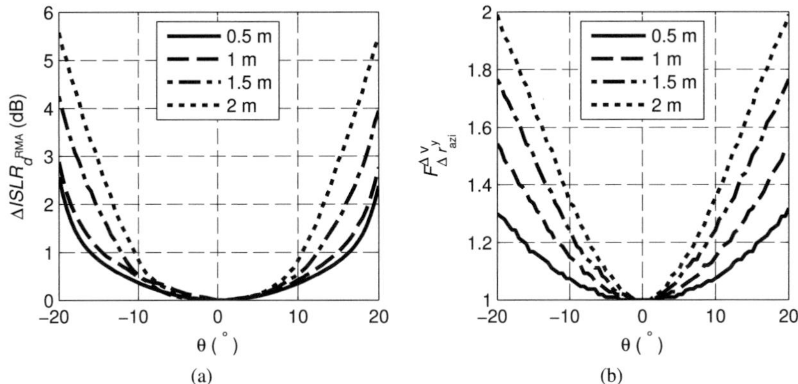

Figure 3.18: The influences of Δv_y in the cases of $f_c = 24\,\text{GHz}$, $\Psi = 40°$ and $y'_t = 0.5, 1, 1.5$ and $2\,\text{m}$: (a) increasing $ISLR_{d\text{RMA}}$; (b) broadening $\Delta r_{\text{azi,eff}}$.

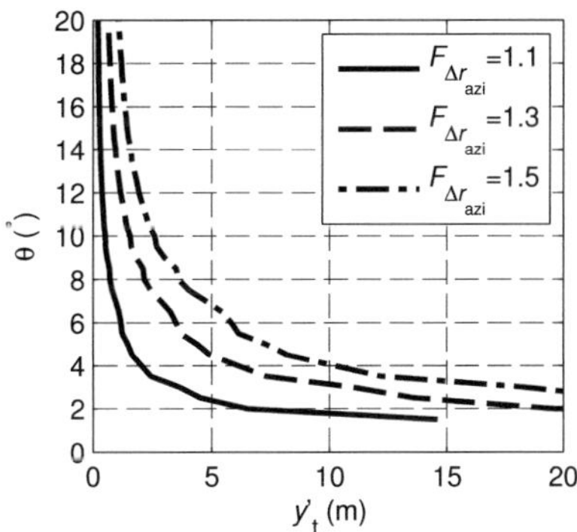

Figure 3.19: The relationships between maximum allowable θ and y'_t at different allowable broadening factors, $F_{\Delta r_{\text{azi}}}^{\Delta v_y}$.

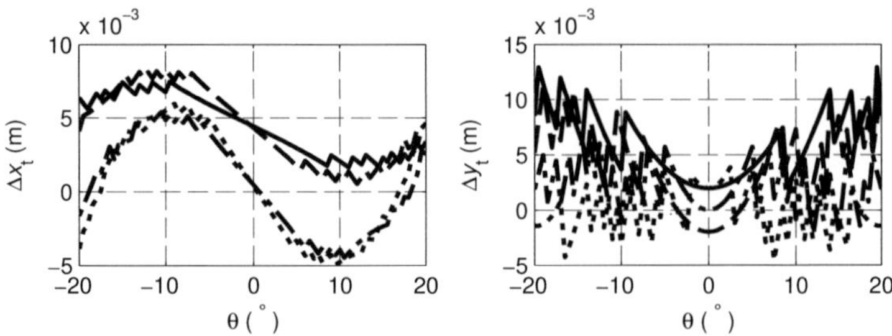

Figure 3.20: Shifts caused by Δv_y along azimuth and range directions, where results of $y'_t = 0.5, 1, 1.5$ and $2\,$m are plotted with line types of solid, dashed, dash-dot and dotted respectively.

results from the quantized simulations. As indicated in (3.57), the phase error in spatial frequency domain is linearly proportional to y'_t, which results in the same relationship in the space domain. It is seen that for a certain θ, the interval between the broadening factor is also linearly proportional to the difference between y'_t with a coefficient depending on θ, such relationship can be expressed as

$$F_{\Delta r_{azi}}^{\Delta v_y}\left(y'_t, \theta\right) = K_1 \cdot y'_t \cdot f\left(\theta\right) + 1, \tag{3.62}$$

where K_1 is a constant coefficient, and $f\left(\theta\right)$ is a function of θ which is zero when $\theta = 0$. Since it is difficult to estimate the form of $f\left(\theta\right)$, a 2D curve fitting for $F_{\Delta r_{azi}}^{\Delta v_y}$ using the simulation results cannot be performed. However, a series of linear functions in the form of $F_{\Delta r_{azi}}^{\Delta v_y}\left(y'_t, \theta\right) = K_\theta \cdot y'_t + 1$ at different θ can be obtained by curve fitting to a degree 1 polynomial. Given these functions, the relationship between the maximum allowable θ for a certain y'_t for a maximum allowable $F_{\Delta r_{azi}}^{\Delta v_y}$ can be obtained, and the results are shown in Figure 3.19.

The coordinates of the strongest point, denoted with $(\tilde{x}_t, \tilde{y}_t)$, in the SAR image have been determined and the difference, denoted with $(\Delta x_t, \Delta y_t)$, between $(\tilde{x}_t, \tilde{y}_t)$ and the nominal coordinates, (x_t, y_t), have been calculated, and the results are shown in Figure 3.20. Besides the obvious constant difference between two groups of y'_t which is caused by the quantized simulations, the shift along azimuth within about $|\theta| < 10°$ is linear, which is the effect of the quasi-cubic phase error of (3.57) as discussed in Appendix B. Beyond the linear region, Δx_t changes like a sinusoid function. The reason for this irregular shift error may be as follows. As the shift of the spectrum along k_x axis becomes larger, the region of the support band is deformed as illustrated in Figure 3.16a. Thus the IRF of the SAR system is deformed too, which results in this irregular shift. The phase error in (3.57) also shifts the target along range direction as illustrated in the right-hand side of Figure 3.20. The obvious two constant differences between three groups of y'_t are also caused by the quantized simulations. The irregular change of Δy_t results from the deformed support band as in the case of Δx_t. Nevertheless, these irregular changes of $(\Delta x_t, \Delta y_t)$ are relatively small compared to the system's resolution in azimuth and range directions and thus can be ignored.

The above analysis assumes the nominal velocity, v, is used for SAR processing. However, in practical applications the real velocity, v', is actually used. There are two reasons for this operation. First, according to (3.51) v is a function of the instantaneous θ which, though it is constant for a certain near range target, changes during the whole SAR processing block. Second, by analogy with the case of Δy_0, the SAR data after first-order motion compensation for Δv_y is intrinsically equivalent to the SAR data containing the "residual" linear motion error expressed in the $x'y'$-coordinate system as

$$
\begin{aligned}
\tilde{s}'_{\mathrm{IF}}\left(x'_{\mathrm{am}}, k_r\right) &= s'_{\mathrm{IF}}\left(x'_{\mathrm{am}}, k_r\right) \cdot \exp\left(-jk_r \cdot \Delta v_y t\right) \\
&= \tilde{s}_{\mathrm{IF}}\left(x_{\mathrm{am}}, k_r\right) \cdot \exp\left(-jk_r \cdot \Delta v_y t\right) \\
&= s_{\mathrm{IF}}\left(x_{\mathrm{am}}, k_r\right) \cdot \exp\left[-jk_r \cdot \Delta r_{\mathrm{var}}\left(\Delta v_y, x_t, y_t\right)\right].
\end{aligned}
\tag{3.63}
$$

In this case, $\Delta v_y t$, is unknown to the user and v' is used instead during the SAR processing.

Besides rotating the target's coordinates from (x'_t, y'_t) to (x_t, y_t) via (3.54) and (3.55) and deteriorating the IRF of the system as illustrated in Figure 3.18, further deterioration will be caused as a result of using v' instead of v. The resulting constant velocity error in the azimuth direction can be calculated as

$$
\Delta v_x = v' - v = v \cdot \left(\frac{1}{\cos\theta} - 1\right).
\tag{3.64}
$$

As analyzed in subsection 3.3.2, the influences of Δv_x include a shift and defocus in azimuth direction.

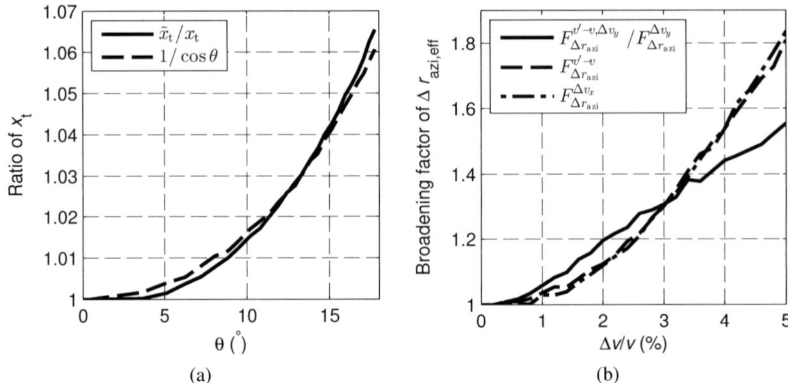

Figure 3.21: Influences of using v' instead of v in the case of $f_c = 24\,\text{GHz}$, $y_t = 1\,\text{m}$ and $\Psi = 40°$: (a) shift along azimuth; (b) broadening of $\Delta r_{\text{azi,eff}}$, where $\frac{\Delta v_x}{v} = 5\,\%$ corresponds to about $17.8°$.

First, substituting (3.54) and (3.64) into (3.16), the shifted position in x' axis can be calculated as

$$
\begin{aligned}
\tilde{x}_t &= x_t + x_t \cdot \frac{\Delta v_x}{v} \\
&= \left(x_t' \cdot \cos\theta - y_t' \cdot \sin\theta \right) \cdot \frac{1}{\cos\theta} \\
&= x_t' - y_t' \cdot \tan\theta,
\end{aligned}
\tag{3.65}
$$

which is in agreement with [60], where the difference between v' and v is overlooked. As shown in Figure 3.21a, the ratio of $\tilde{x}_t$ to x_t determined from simulations is approximately in accordance with $\frac{1}{\cos\theta}$ as indicated in (3.65). The difference between these two lines comes from the shift caused by Δv_y as illustrated in Figure 3.20, and should be disappeared in the case of small Ψ where the above discussed space-variant motion error is trivial as in [60].

Second, Δv_x also causes defocus in azimuth direction on the basis of the defocused IRF in the presence of space-variant motion errors as discussed above, or on the basis of an ideal IRF as in [60] where narrow Ψ is presumed. In Figure 3.21b, $F_{\Delta r_{\text{azi}}}^{v'-v,\Delta v_y}$ is the broadening factor of Δr_{azi} determined with v' in the presence of space-variant motion errors and $F_{\Delta r_{\mathrm{azi}}}^{\Delta v_y}$ is the broadening factor depicted in Figure 3.18b; $F_{\Delta r_{\mathrm{azi}}}^{v'-v}$ is the ratio of Δr_{azi} measured with v' to Δr_{azi} measured with v with the space-variant motion errors being compensated by a second-order motion compensation, which are intrinsically equivalent to the cases of narrow Ψ; $F_{\Delta r_{\mathrm{azi}}}^{\Delta v_x}$ is the result obtained in subsection 3.3.2 where only Δv_x is considered and plotted here as a reference. It is seen that the dashed line and the dash-dot line are almost overlapped, which means that without the influences of space-variant motion errors as in the case of narrow Ψ, using v' instead of v causes the same broadening effect on Δr_{azi} as analyzed in subsection 3.3.2. On the other hand, if the space-variant motion errors are not compensated, it is seen that $F_{\Delta r_{\mathrm{azi}}}^{v'-v,\Delta v_y}/F_{\Delta r_{\mathrm{azi}}}^{\Delta v_y}$, which reflects the broadening effect caused by $v'-v$, increases almost linearly but with a smaller slope than that of the reference $F_{\Delta r_{\mathrm{azi}}}^{\Delta v_x}$. As discussed in subsection 3.3.2, the quadratic increasing style of $F_{\Delta r_{\mathrm{azi}}}^{\Delta v_x}$ within the region of small $\frac{\Delta v_x}{v}$ is caused by the annular shape of the support band in the spatial frequency domain. Such effect has also been included when measured $F_{\Delta r_{\mathrm{azi}}}^{\Delta v_y}$ as illustrated in Figure 3.18b. Therefore this non-linear increasing style of $F_{\Delta r_{\mathrm{azi}}}$ is canceled out when calculating $F_{\Delta r_{\mathrm{azi}}}^{v'-v,\Delta v_y}/F_{\Delta r_{\mathrm{azi}}}^{\Delta v_y}$. Though the slope of $F_{\Delta r_{\mathrm{azi}}}^{v'-v,\Delta v_y}/F_{\Delta r_{\mathrm{azi}}}^{\Delta v_y}$ is smaller than that of $F_{\Delta r_{\mathrm{azi}}}^{\Delta v_x}$, the total broadening factor $F_{\Delta r_{\mathrm{azi}}}^{v'-v,\Delta v_y}$ taking into account both space-variant motion errors, $\Delta r_{\mathrm{var}}(\Delta v_y, x_{\mathrm{t}}, y_{\mathrm{t}})$, and constant nominal velocity error, $\Delta v_x = v'-v$, can be calculated approximately by multiplying $F_{\Delta r_{\mathrm{azi}}}^{\Delta v_x}$ in Figure 3.9b and $F_{\Delta r_{\mathrm{azi}}}^{\Delta v_y}$ in Figure 3.18b, since the phase errors related to both of them have been analytically derived and obtained as (3.24) and (3.57).

3.4.3 Quadratic Motion Error

Like the case of Δv_y, though the influence of QPE in the spatial frequency domain is well known in the literature [18], the link between the quadratic motion error, $\frac{1}{2}\Delta a_y t^2$, and QPE in the spectrum was not yet derived for stripmap SAR with wide Ψ. To quantitatively analyze the influence of $\frac{1}{2}\Delta a_y t^2$, the relationship between the motion parameters and the resulting phase error in the spatial frequency domain is derived in the following.

First of all the $\frac{1}{2}\Delta a_y t^2$ deteriorated IF data can be expressed as

$$\tilde{s}_{\mathrm{IF}}(x_{\mathrm{am}}, k_r) = A_t w_r(k_r) w_a(\tilde{r}_t) \exp\left[-jk_r \cdot \sqrt{(x_{\mathrm{am}} + v\hat{i} - x_t)^2 + \left(\frac{1}{2}\Delta a_y t^2 - y_t\right)^2}\right].$$

(3.66)

Since no antenna servo is assumed in this dissertation, a quadratic trajectory causes the increase or decrease of Doppler bandwidth, B_{Dop}, via the term $w_a(\tilde{r}_t)$. The increased B_{Dop} may cause azimuth aliasing when the *PRF* cannot meet the Nyquist criterion, while the decreased B_{Dop} will broaden Δr_{azi} as dictated by (2.25). These effects are investigated via simulations together with those caused by phase errors.

The motion errors for target t, $\Delta r_t(\Delta a_y)$, caused by $\frac{1}{2}\Delta a_y t^2$ can be calculated by substituting $\Delta v_y t$ with $\frac{1}{2}\Delta a_y t^2$ into (3.56) and that into (3.3) as

$$\begin{aligned}
\Delta r_t(\Delta a_y) &= \Delta r_{\mathrm{var}} + \Delta r_{\mathrm{in}} \\
&= \frac{\frac{1}{2}\Delta a_y t^2}{2\left(y_t - \frac{1}{2}\Delta a_y t^2\right) y_t} \cdot (x_{\mathrm{am}} + v\hat{i} - x_t)^2 - \frac{1}{2}\Delta a_y t^2.
\end{aligned}$$

(3.67)

According to POSP, by substituting $x_{\mathrm{am}} + v\hat{i} - x_t$ with $-y_t \frac{k_x}{k_y}$, and $t = -\frac{y_t}{v} \cdot \frac{k_x}{k_y}$ into (3.67), the phase error in the spatial frequency domain can be obtained as

$$\exp\left[-jk_r \Delta r_t(\Delta a_y)\right] = \exp\left[jk_r \cdot \left(1 - \frac{\tan^2 \psi}{2 - \Delta a_y \cdot \frac{y_t}{v^2} \cdot \tan^2 \psi}\right) \cdot \frac{1}{2}\Delta a_y \left(\frac{y_t}{v}\right)^2 \tan^2 \psi\right],$$

(3.68)

with $\tan \psi = k_x/k_y$. Some comments on (3.68) are now in order. First, both (3.67) and (3.68) are derived based on the assumption that $y_t - \frac{1}{2}\Delta a_y t^2 \gg |x_{\mathrm{am}} + v\hat{i} - x_t| > 0$, so the term in the brackets can be approximated to 1 when investigating the effect of Δr_t. Second, it is seen that the phase error caused by $\frac{1}{2}\Delta a_y t^2$ is linearly proportional to, $k_r \cdot \tan^2 \psi$, which results in broadening of IRF in azimuth direction. Therefore, if the optimum Ψ obtained in subsection 2.4.3 is used, the ratio between the degrees of the influence caused by the same quadratic motion error for $f_c = 24, 77$ and $120\,\mathrm{GHz}$ is $1 : 0.75 : 1.17$.

In order to obtain a realistic range of the phase error in (3.68), ISO 2631 is used analogously to the cases of Δy_0 and Δv_y as follows.

1. A quadratic motion error can be represented by a sinusoidal motion error with $MSR = 0.5$. Then according to (3.6), the vibration frequency of the sinusoid can be calculated as

$$
\begin{aligned}
f_y &= MSR \cdot \frac{v}{L_t} \\
&= \frac{1}{4\tan\frac{\Psi}{2}} \cdot \frac{v}{y_t}.
\end{aligned}
\tag{3.69}
$$

In the case of $f_c = 24\,\text{GHz}$ with optimum $\Psi = 40°$ as designated in subsection 2.4.3, the range of $f_y \in [0.04, 34.34]\,\text{Hz}$ can be obtained given the range of $v \in [1, 25]\,\text{m/s}$ and $y_t \in [0.5, 16]\,\text{m}$. Subsequently the corresponding values of $W_d(f_y)$ can be determined as described in Appendix A.

2. By analogy with the analysis in subsection 3.3.4, 5 different DoD are given with a_v being $0.4725\,\text{m/s}^2$, $0.75\,\text{m/s}^2$, $1.2\,\text{m/s}^2$, $1.875\,\text{m/s}^2$ and $2\,\text{m/s}^2$ for DoD of 2, 3, 4 and 5 respectively. Accordingly, 5 series of A_a corresponding to the same range of f_y specified in step 1 can be obtained via (A.7).

3. Since a quadratic motion error, $\frac{1}{2}\Delta a_y t^2$, approximates to a sinusoid of half period, the amplitude of the sinusoid equals to the quadratic motion error at the edge of the synthetic aperture, i.e.

$$
\frac{A_a}{(2\pi f_y)^2} = \frac{1}{2}\Delta a_y \left(\frac{L_t}{2v}\right)^2 = \frac{1}{2}\Delta a_y \left(\frac{y_t}{v}\right)^2 \tan^2\frac{\Psi}{2}.
\tag{3.70}
$$

Substituting (3.69) into (3.70), Δa_y can be calculated as

$$
\Delta a_y = \frac{8A_a}{\pi^2}.
\tag{3.71}
$$

As indicated in (A.7), A_a is a function of f_y, and hence Δa_y depends on f_y too. Given the values of A_a specified in step 2, the relationships between Δa_y and f_y for different DoD can be obtained, which are plotted in Figure 3.22. In addition, by analogy with the analyses of Δy_0 and Δv_y, some values of Δa_y are restricted for $f_y < 0.2\,\text{Hz}$, the criterion is that the magnitude of the quadratic motion error

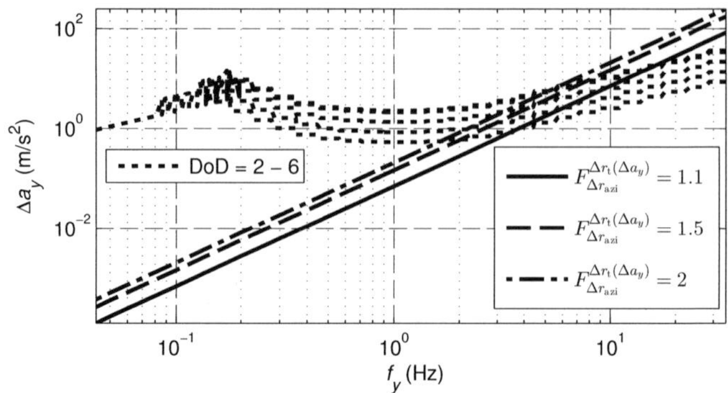

Figure 3.22: Relationships between Δa_y, f_y for a series of DoD (each short-dashed curve represents a DoD with the uppermost curve corresponding to DoD $= 6$ and the lowest curve corresponding to DoD $= 2$), and relationships between f_y, DoD and $F_{\Delta r_{azi}}^{\Delta r_t(\Delta a_y)}$, illustrated through the 3 curves representing different $F_{\Delta r_{azi}}^{\Delta r_t(\Delta a_y)}$ values and the 5 curves representing different DoD.

should not exceed y_t, i.e.

$$\frac{A_a}{(2\pi f_y)^2} = \frac{\Delta a_y \cdot \frac{\pi^2}{8}}{(2\pi f_y)^2} < y_t. \tag{3.72}$$

First, a series of simulations have been performed with $y_t = 1\,\mathrm{m}$ and $v = 1\,\mathrm{m/s}$ for different values of Δa_y based on the results illustrated in Figure 3.22. As illustrated in Figure 3.23a, $\Delta ISLR_{d\mathrm{RMA}}$ caused by Δa_y is not symmetric about $\Delta a_y = 0$, but grows faster when $\Delta a_y > 0$ than when $\Delta a_y < 0$. The primary reason is the aliasing effect resulting from the increased B_{Dop} via the antenna-beam-pattern-related $w_a(\tilde{r}_t)$ when $\Delta a_y > 0$. On the contrary, the contribution to such asymmetry from the space-variant component of Δr_t as indicated in (3.68) can be ignored. The asymmetry is also seen in $F_{\Delta r_{azi}}^{\Delta r_t(\Delta a_y)}$ as a function of Δa_y in Figure 3.23b. Therefore the allowable levels of the magnitude of Δa_y should use the values of $\Delta a_y > 0$. It should be noted that, as will be explained later, the change of B_{Dop} is inversely proportional to y_t given the same Δa_y,

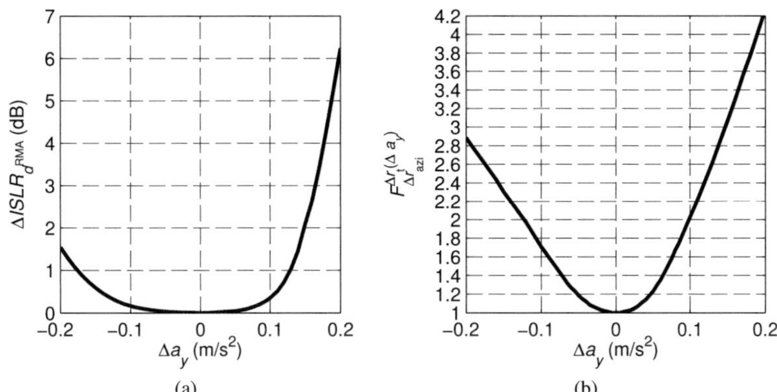

(a)   (b)

Figure 3.23: The influences of $\Delta r_t\,(\Delta a_y)$ in the cases of $f_c = 24\,\text{GHz}$, $\Psi = 40°$, $y_t = 1\,\text{m}$ and $v = 1\,\text{m/s}$: (a) increasing $ISLR_{d\text{RMA}}$; (b) broadening $\Delta r_{\text{azi,eff}}$.

i.e. the same phase error at the edge of the synthetic aperture. Therefore the results shown in Figure 3.23 can be considered as the worst case.

In addition, when the prerequisite of the deduction of (3.68) is fulfilled, for a given Ψ and within the region of $|\Delta a_y| < 0.2\,\text{m/s}^2$, the phase of (3.68) is approximately linearly proportional to the term $\Delta a_y \cdot \frac{y_t^2}{v^2}$. Therefore, the results obtained for $\frac{y_t}{v} = 1$ as shown in Figure 3.23b can be used to deduce for other combinations of y_t and v, which are related to f_y as indicated in (3.69). For instance, given the values of Δa_y in the case of $\frac{y_t}{v} = 1$, the values of Δa_y for other $\frac{y_t}{v} = \frac{1}{4\tan\frac{\Psi}{2} f_y}$ can be calculated as a function of f_y as

$$\Delta a_y\,(f_y) = \Delta a_y\Big|_{\frac{y_t}{v}=1} \cdot \frac{\left(\frac{y_t}{v}\right)^2\Big|_{\frac{y_t}{v}=1}}{\left(\frac{y_t}{v}\right)^2} = \Delta a_y\Big|_{\frac{y_t}{v}=1} \cdot \frac{f_y^2}{f_y^2\big|_{\frac{y_t}{v}=1}}. \tag{3.73}$$

Three lines of Δa_y as a function of f_y corresponding to $F_{\Delta r_{\text{azi}}}^{\Delta r_t(\Delta a_y)} = 1.1,\ 1.5$ and 2 are plotted in Figure 3.22. It is seen that for $f_y > 11\,\text{Hz}$, which corresponds to high v or small y_t or both, even the vibration with $\text{DoD} = 6$ will not cause severe deterioration on the SAR image. On the contrary, for $f_y < 1.1\,\text{Hz}$, e.g. $f_y = \frac{1}{4\tan(20°)} \cdot \frac{16\,\text{m/s}}{10\,\text{m}} \approx 1.1\,\text{Hz}$, even the vibration of $\text{DoD} = 2$ can result severe broadening of Δr_{azi}. Thus, it is necessary to apply motion compensation on the SAR data for automotive applications.

In the following paragraphs the effects of space-variant motion error, $\Delta r_{\text{var}}(\Delta a_y)$, are considered. After first-order motion compensation, the phase error caused by Δr_{var} in the spatial frequency domain can be obtained from (3.68) as

$$\exp\left[-jk_r\Delta r_{\text{var}}(\Delta a_y)\right] = \exp\left(-jk_r \cdot \frac{\Delta a_y \frac{y_t^2}{v^2}\tan^4\psi}{4 - 2\Delta a_y \cdot \frac{y_t}{v^2}\cdot \tan^2\psi}\right). \qquad (3.74)$$

As will be illustrated later, the denominator term, $4 - 2\Delta a_y \cdot \frac{y_t}{v^2}\cdot \tan^2\psi$, in (3.74) is almost constant within the the range of Δa_y that is considered, i.e. the prerequisite of the deductions of (3.67) and (3.68) is fulfilled. Therefore, (3.74) is a quasi-quartic function of k_x. However, there is no easy form of closed expression corresponding to (3.74) in space domain after performing 2D IFFT. Furthermore, in the case of large $|\Delta a_y|$ the influence of the change of B_{Dop} dictated by the term $w_a(\tilde{r}_t)$ in (3.66) may become dominant over that of the phase error dictated by (3.68). Therefore, to investigate the influence of Δr_{var}, a first-order motion compensation have been applied on the SAR data set with $y_t = 1, 2, 4$ and $8\,\text{m}$ respectively, and the results of $F_{\Delta r_{\text{azi}}}^{\Delta r_{\text{var}}(\Delta a_y)}$ and $\Delta ISLR_{d\text{RMA}}$ as functions of Δa_y are illustrated in Figure 3.24.

Some comments are now in order.

First, according to the simulation results, deterioration caused both by phase error and change of B_{Dop} is linearly proportional to the multiplication term, $\frac{1}{2}\Delta a_y\left(\frac{y_t}{v}\right)^2\tan^2\frac{\psi}{2}$, i.e. the quadratic motion error at the edge of synthetic aperture. Therefore, the horizontal axis of each figure is set to be $\frac{1}{2}\Delta a_y\left(\frac{y_t}{v}\right)^2\tan^2\frac{\psi}{2}$, and the allowable levels of Δa_y at different v for a certain y_t can be determined accordingly.

Second, for $\Delta a_y < 0$, it is seen that in the cases of $y_t = 1\,\text{m}$ and $y_t = 2\,\text{m}$, it is the decrease of B_{Dop} that dominantly broadens Δr_{azi}, which does not not degrade $ISLR_{d\text{RMA}}$. In the case of $y_t = 4\,\text{m}$, within the range of $\Delta a_y \in [-3,0]\,\text{m/s}^2$, the influences of phase error, i.e. increasing both $F_{\Delta r_{\text{azi}}}^{\Delta r_{\text{var}}(\Delta a_y)}$ and $\Delta ISLR_{d\text{RMA}}$, are dominant over that of the decreased B_{Dop}. While within the range of $\Delta a_y \in [-10,-3]\,\text{m/s}^2$, the decrease of B_{Dop} becomes the primary cause of the deterioration, whereas $\Delta ISLR_{d\text{RMA}}$ keeps almost constant. In the case of $y_t = 8\,\text{m}$, the critical point of the change of primary cause of deterioration is at approximate $\Delta a_y = -5\,\text{m/s}^2$. The reason for the different critical point for different y_t is as follows: for the same $\frac{1}{2}\Delta a_y\left(y_t\tan\frac{\psi}{2}/v\right)^2$, the smaller the y_t, the bigger the looking angle at the edge of the synthetic aperture with regard to the center line of the antenna beam pattern, which causes a more severe decrease of B_{Dop}. Furthermore, it is seen that the slopes of $F_{\Delta r_{\mathrm{azi}}}^{\Delta r_{\mathrm{var}}(\Delta a_y)}$ for $y_t = 4\,\mathrm{m}$ and $y_t = 8\,\mathrm{m}$ are almost the same, so the slope of $F_{\Delta r_{\mathrm{azi}}}^{\Delta r_{\mathrm{var}}(\Delta a_y)}$ for $y_t = 8\,\mathrm{m}$ can be used in the case of $y_t > 8\,\mathrm{m}$ for determining the effect of $\Delta r_{\mathrm{var}}(\Delta a_y)$.

Third, for $\Delta a_y > 0$, it is seen that the primary cause of increased $\Delta ISLR_{d\mathrm{RMA}}$ changes from increase of B_{Dop} for near-range targets to phase error for far-range targets, while the slope of $F_{\Delta r_{\mathrm{azi}}}^{\Delta r_{\mathrm{var}}(\Delta a_y)}$ keeps approximately the same for any y_t.

Fourth, since the effect of $\frac{1}{2}\Delta a_y \left(y_t \tan \frac{\Psi}{2}/v\right)^2$ is inversely proportional to y_t taking into account the effect of rotation as illustrated in Figure 3.24, the allowable levels of quadratic motion error (at the edge of synthetic aperture), $\frac{1}{2}\Delta a_y \left(y_t \tan \frac{\Psi}{2}/v\right)^2$, when only first-order motion compensation is applied, should be determined first according to the results of furthest target of interest, and then deduced for nearer targets. Furthermore, as discussed in the second comment, the results of furthest target can be approximated by the results of $y_t = 8\,\mathrm{m}$. In the case of $f_c = 24\,\mathrm{GHz}$ and $\Psi = 40^\circ$, $F_{\Delta r_{\mathrm{azi}}}^{\Delta r_{\mathrm{var}}(\Delta a_y)}$ and $\Delta ISLR_{d\mathrm{RMA}}$ as functions of $\frac{1}{2}\Delta a_y \left(y_t \tan \frac{\Psi}{2}/v\right)^2$ and the corresponding QPE for $y_t > 8\,\mathrm{m}$ are illustrated in Figure 3.25. It is seen that with first-order motion compensation, the tolerable QPE, for $F_{\Delta r_{\mathrm{azi}}}^{\Delta r_{\mathrm{var}}(\Delta a_y)} = 1.2$, is approximate $\pm 40 \cdot \pi$, which is quite an improvement comparing with the result without compensation, e.g. the tolerable QPE for $F_{\Delta r_{\mathrm{azi}}}^{\Delta r_{\mathrm{var}}(\Delta a_y)} = 1.5$ is π as presented in [18]. On the one hand, given a required $F_{\Delta r_{\mathrm{azi}}}^{\Delta r_{\mathrm{var}}(\Delta a_y)}$ at certain v for application of certain $y_{t,\mathrm{max}}$, the allowable level of Δa_y can be deduced from Figure 3.25. On the other hand, for a certain allowable $\frac{1}{2}\Delta a_y \left(y_t \tan \frac{\Psi}{2}/v\right)^2$ dictated according to Figure 3.25, e.g. which results in $F_{\Delta r_{\mathrm{azi}}}^{\Delta r_{\mathrm{var}}(\Delta a_y)} = 1.2$, its influence on targets of $y_t < 8\,\mathrm{m}$ is inversely proportional to y_t^2, thus resulting in $F_{\Delta r_{\mathrm{azi}}}^{\Delta r_{\mathrm{var}}(\Delta a_y)} < 1.2$.

3.4.4 Sinusoidal Motion Error

As already discussed in [18], a sinusoidal phase error in spatial frequency domain causes high paired sidelobes in the final SAR image. The positions and magnitudes of the sidelobes depend on the frequency and amplitude of the sinusoidal phase error, which has been analytically derived in [18].

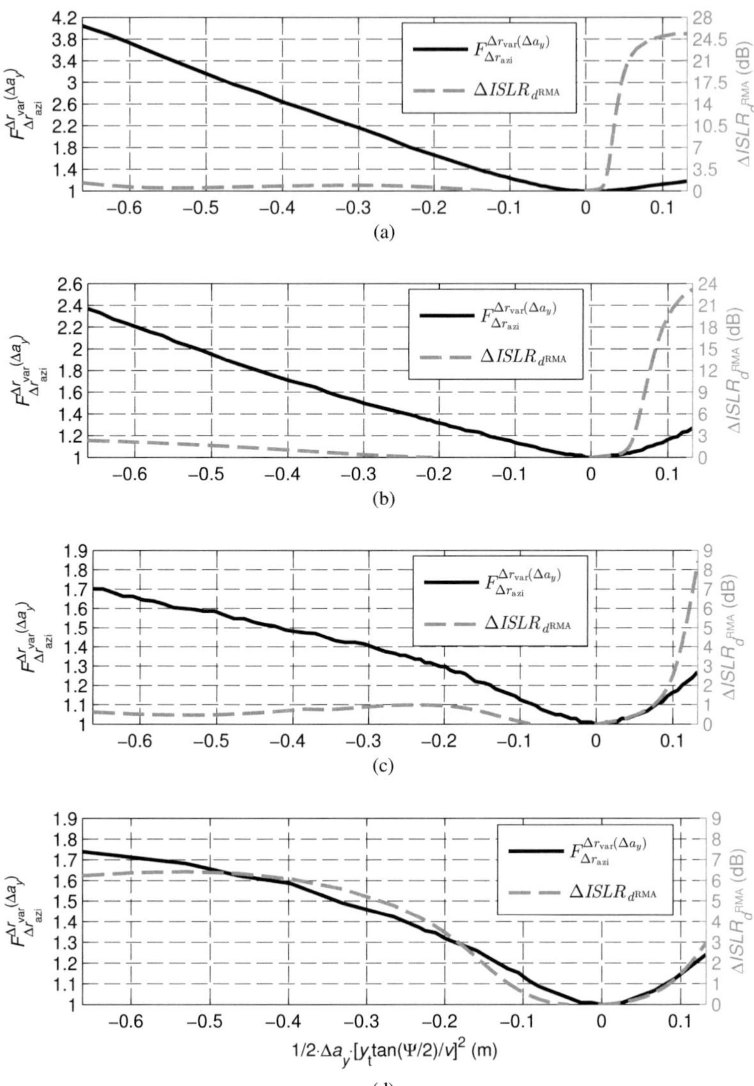

Figure 3.24: The influences of $\Delta r_{\mathrm{var}}(\Delta a_y)$ in the cases of $f_{\mathrm{c}} = 24\,\mathrm{GHz}$ and $\Psi = 40°$ as functions of the quadratic motion error at the edge of the synthetic aperture: (a) $y_{\mathrm{t}} = 1\,\mathrm{m}$; (b) $y_{\mathrm{t}} = 2\,\mathrm{m}$; (c) $y_{\mathrm{t}} = 4\,\mathrm{m}$; (d) $y_{\mathrm{t}} = 8\,\mathrm{m}$.

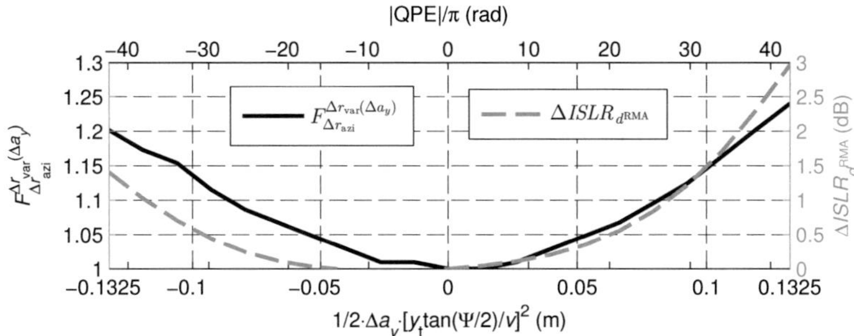

Figure 3.25: The broadening factor, $F_{\Delta r_{azi}}^{\Delta r_{var}(\Delta a_y)}$, and $\Delta ISLR_{d\mathrm{RMA}}$ as functions of the quadratic motion error (or the corresponding QPE) at the edge of the synthetic aperture in the cases of $f_c = 24\,\mathrm{GHz}$ and $\Psi = 40°$ after applying the first-order motion compensation.

According to POSP the linear mapping from a sinusoidal motion error, $A_y \cdot \sin(2\pi f_y t)$, to a sinusoidal phase error in spatial frequency only exists when the fluctuations of $A_y \cdot \sin(2\pi f_y t)$ along azimuth direction are ignorable compared to that of the raw SAR signal, (2.43). Moreover relatively high fluctuations of $A_y \cdot \sin(2\pi f_y t)$ also imply high fluctuations of the azimuth weighting function, $w_a(\tilde{r}_t)$, which results in an aliasing effect in the same way as in the case of quadratic motion error.

The necessity of compensating $A_y \cdot \sin(2\pi f_y t)$ has been proven indirectly in subsection 3.4.3. The quadratic motion error for $y_t = 1\,\mathrm{m}$ with a certain DoD that has been shown requiring compensation is equivalent to a sinusoidal motion error for $y_t > 2\,\mathrm{m}$.

3.5 Vertical Motion Errors

The effects of a vertical motion error, Δz, are the same as that of a range motion error, Δy, but with much less magnitude. As illustrated in Figure 3.26, for the targets located in the same plane with the antenna's nominal LOS, the slant range motion error caused by a vertical motion error, $\Delta r_t(\Delta z)$, is much smaller than that caused by a range motion error, $\Delta r_t(\Delta y)$, given $\Delta y = \Delta z$, and the relationship between them can be approximated

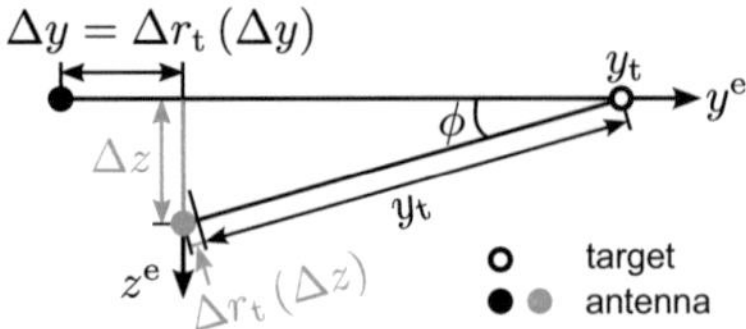

Figure 3.26: Slant range error, Δr_{t}, caused by Δz and Δy.

as

$$\Delta r_{\mathrm{t}}\left(\Delta z\right) \approx \Delta r_{\mathrm{t}}\left(\Delta y\right)\big|_{\Delta y=\Delta z}\cdot\phi. \tag{3.75}$$

Here, ϕ is the angle between the slant range with vertical motion error and the IDP, i.e. the horizontal plane in this case, and therefore $\Delta r_{\mathrm{t}}\left(\Delta z\right)$ is also range dependent.

Moreover for targets located outside of the same horizontal plane with the radar sensor, compensating for $\Delta r_{\mathrm{t}}\left(\Delta z\right)$ calculated under the assumption that targets are in the same height level with antenna will introduce extra phase error which may result in a worse SAR image than the case without compensation. Therefore although the influences and the allowable levels of Δz can be obtained simply from the results obtained in section 3.4, it is not taken into account in the following discussion of motion compensation.

4 Range Motion Compensation Algorithm

As discussed in section 3.4, for near-range SAR systems equipped with a wide beam-width antenna, performing a first-order motion compensation may introduce artificial motion errors into the near range targets signals. As result, near range targets will suffer from loss of relative position information and azimuth resolution degradation. This is, in principle, a problem of space-variant motion compensation. Besides the four algorithms introduced in Chapter 1, a new motion compensation algorithm for stripmap mode SAR is proposed in section 4.1, which exclusively exploits the feature of the geometry of near-range SAR applications.

In section 4.2, the other aspect of motion compensation, i.e. motion measurement, is discussed. This work focuses on measuring slant range motion errors by a hardware-based approach. Generally, at least 6 degrees of freedom (DoF) are required to form a complete inertial navigation system (INS) to provide the coordinates of the system in the Earth-fixed frame. For motion compensation of near-range SAR applications, very high accuracy is not necessarily the most important issue, sometimes it is compromised to meet performance specifications at reduced cost and size. By exploiting the feature of near-range SAR applications, two algorithms using low-cost MEMS IMU sensors of less than 6 DoF to obtain the range coordinates of radar sensor are proposed, and the corresponding requirements for motion sensors are analyzed quantitatively.

4.1 Octave Division Motion Compensation Algorithm

Assuming the motion data of the radar sensors are correctly obtained, the block diagram of octave division motion compensation is shown in Figure 4.1.

First of all, the raw SAR IF signal deteriorated by motion errors are transferred from range-frequency azimuth-time domain, $\tilde{s}_{\mathrm{IF}}(x_{\mathrm{am}}, k_r)$, to range-time azimuth-time domain, $\tilde{s}_{\mathrm{IF,rtat}}(x_{\mathrm{am}}, \hat{t})$, through performing the range IFFT. Then, as illustrated in

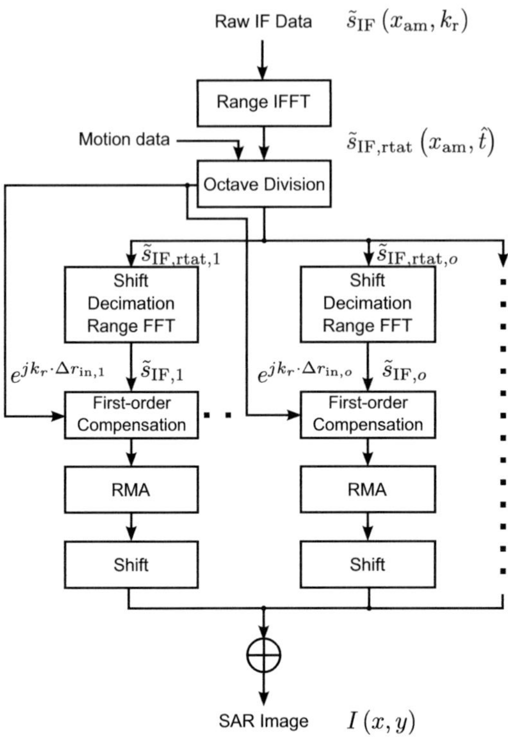

Figure 4.1: Block diagram of octave division motion compensation algorithm.

Figure 4.2, $\tilde{s}_{\mathrm{IF,rtat}}\left(x_{\mathrm{am}}, \hat{t}\right)$ are divided into several overlapped segments both in azimuth and range directions, and the signal of each segment is represented by $\tilde{s}_{\mathrm{IF,rtat},o}$ with subscript "o" denoting the corresponding index. Subsequently, each $\tilde{s}_{\mathrm{IF,rtat},o}$ is shifted to range zero, and since the maximum range of each segment is reduced, the sampling frequency along range direction can be reduced accordingly as dictated by (2.23). The data are then decimated and transferred back to the range-frequency azimuth-time domain, $\tilde{s}_{\mathrm{IF},o}$. To perform the first-order motion compensation, each $\tilde{s}_{\mathrm{IF},o}$ is multiplied with the corresponding motion compensation phase term, $\exp\left(jk_r \cdot \Delta r_{\mathrm{in},o}\right)$, where the nominal trajectory for each segment is obtained by performing linear fitting on the corresponding motion data. As will be explained in the following subsection, the principle of octave division guarantees that artificial motion errors will be excluded almost

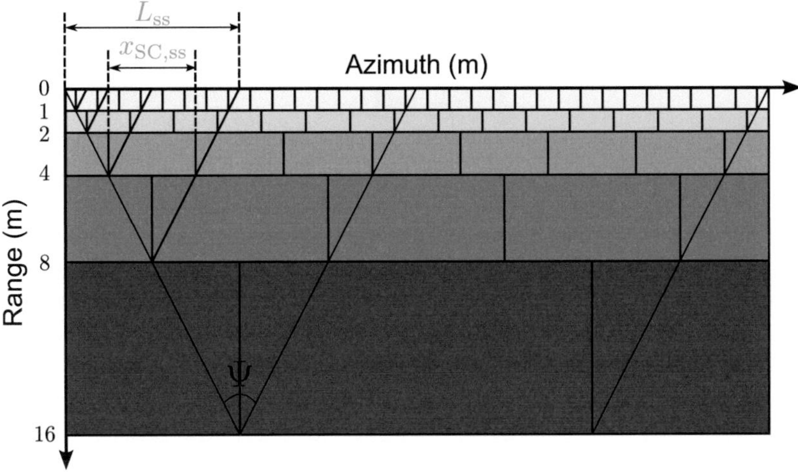

Figure 4.2: Octave division in data space $\tilde{s}_{\mathrm{IF,rtat}}(x_{\mathrm{am}},\hat{t})$, where L_{ss} and $x_{\mathrm{SC,ss}}$ of subswath of $y \in [2,4]$ m are denoted.

completely from $\Delta r_{\mathrm{in},o}$. In the end, the result of each segment is shifted both in range and azimuth directions and then added to form the final SAR image, $I(x,y)$.

4.1.1 Division Principle

In this subsection the core and the origin of the name of octave division motion compensation algorithm, i.e. the principle of the octave division, are explained.

As illustrated in Figure 3.6, the artificial motion errors arise from the difference between nominal trajectories with regard to the furthest range and the nearest range targets. The degree of such difference can be assessed by the ratio of the furthest range to the nearest range within the scene to be imaged, FNR, which can be expressed as

$$FNR = \frac{r_{\mathrm{t},0,\mathrm{Furthest}}}{r_{\mathrm{t},0,\mathrm{Nearest}}}. \tag{4.1}$$

By substituting (3.5) and (3.6) into (4.1), *FNR* can be expressed as the ratio of *MSR* of the furthest target to *MSR* of the nearest target as

$$FNR = \frac{\frac{2 \cdot r_{t,0,\text{Furthest}} \cdot \tan\left(\frac{\Psi}{2}\right)}{\frac{v}{f_m}}}{\frac{2 \cdot r_{t,0,\text{Nearest}} \cdot \tan\left(\frac{\Psi}{2}\right)}{\frac{v}{f_m}}} = \frac{MSR_{\text{Furthest}}}{MSR_{\text{Nearest}}}. \tag{4.2}$$

It is reasonable to assume that there is no artificial motion error for the near range targets if for any motion error when MSR_{Nearest} is 0.25 as the upper limit of the linear region, MSR_{Furthest} is smaller than 0.5 as the middle value of the quadratic region. In such case, any linear motion within the synthetic aperture of the nearest target would approximately not be part of the motion errors that need to be compensated for the furthest target. Moreover, the nominal trajectories for each target within such range swath are approximately overlapped with each other. Therefore the upper limit of *FNR* in the case without artificial motion errors can be obtained as

$$FNR = \frac{MSR_{\text{Furthest}}}{MSR_{\text{Nearest}}} < \frac{0.5}{0.25} = 2. \tag{4.3}$$

By analogy with the octave frequency definition in the sound processing field, the whole range swath can be divided into several octave sub-swaths, i.e. the whole range swath is divided at $y = 1, 2, 4, 8$ and $12\,\text{m}$ as illustrated in Figure 4.2.

As a trade-off between processing efficiency and accommodation of artificial motion errors, for each range sub-swath the imaged scene along azimuth direction is divided into equal segments with $x_{\text{SC,ss}} = 2 \cdot r_{t,0,\text{Furthest,ss}} \cdot \tan\left(\frac{\Psi}{2}\right)$, where subscript "ss" denotes sub-swath. By using (2.2) the corresponding data segment length can be calculated as

$$L_{\text{ss}} = 4 \cdot r_{t,0,\text{Furthest,ss}} \cdot \tan\left(\frac{\Psi}{2}\right). \tag{4.4}$$

Consequently, the overlapped length of segment along azimuth direction is $2 \cdot r_{t,0,\text{Furthest,ss}} \cdot \tan\left(\frac{\Psi}{2}\right)$.

Since decimation in range frequency has been performed for each segment, the extra computational burden caused by the overall overlapped segments here is approximately equal to that of the spot-like motion compensation algorithm as introduced in Chapter 1. Furthermore, the octave division motion compensation algorithm is designed specifically to compensate artificial motion errors, it cannot compensate the residual motion errors resulting from quadratic or higher order sinusoidal motion errors after the first-order motion compensation. Though such residual motion errors are usually smaller than that resulting from artificial motion errors for near range targets. Nevertheless, it can be combined with some of other motion compensation algorithms that deal with the azimuth-variant motion errors if it is necessary. For instance, the proposed algorithm can easily be combined with the spot-like motion compensation algorithm within each segment without requiring extra computation cost just by compensating motion errors relative to each segment center instead to the broadside target.

It should be noted that the artificial motion errors are exclusive to near-range SAR applications. For example, the range swath of SAR system in near-range applications is

$$\frac{2D^2}{\lambda_{\min}} < r_{t,0} < r_{\max}. \tag{4.5}$$

Here, $\frac{2D^2}{\lambda_{\min}}$ is the far-field boundary of the antenna as defined in [122], where D ($D > \lambda_{\max}$) is the maximum overall dimension of antenna and $\lambda_{\min}$ is the minimum wavelength of the transmitted chirp; $r_{\max}$ is the maximum working range of the SAR system as defined in subsection 2.3.4. By applying $D = 4.2\,\mathrm{cm}$, $\lambda_{\min} = 1.28\,\mathrm{cm}$ and $r_{\max} = 18\,\mathrm{m}$, the number of octave sub-swaths can be calculated as

$$N_{\mathrm{sub-swath}} = \log_2(FNR) = \log_2\left(\frac{r_{\max}}{\frac{2D^2}{\lambda_{\min}}}\right) \approx 6. \tag{4.6}$$

As a contrast, in remote sensing SAR application, FNR is usually smaller than 2. For instance, the airborne P-3 SAR used for detection of concealed targets under foliage images the scene with range swath of $[5387.5, 6354.9]$ m [51]. The corresponding FNR can be calculated as

$$FNR = \frac{6354.9}{5387.5} \approx 1.2, \tag{4.7}$$

which means the motion errors have almost the same MSR for all the targets within the whole range swath. Therefore it is unnecessary to perform such octave division motion compensation.

Table 4.1: Coordinates of the 4 points targets in the nominal coordinates axes

Index	x (m)	y (m)
1	0.367	0.500
2	0.725	1.500
3	1.459	3.500
4	2.184	0.500

4.1.2 Verification

To verify the proposed algorithm, a scene with 4 point targets, which are numbered according to their x coordinates in ascending order as 1, 2, 3 and 4, has been simulated. The coordinates of the 4 point targets are listed in Table 4.1.

In Figure 4.3a, the SAR trajectory is depicted on the nominal coordinate axes that are defined with regard to a certain far range target , e.g. $y_t = 15\,\mathrm{m}$, so components of Δy_0 and $\Delta v_y \cdot t$ can be observed.

The result without motion compensation is shown in Figure 4.3b. It is seen that target 2 and 3 suffer greatly from the non-artificial errors component of the trajectory and become unrecognizable, while target 1 and 4 remain almost unaffected since the actual SAR trajectories within their synthetic aperture length are almost linear. Furthermore the slightly shifted positions of target 1 and 4 in the nominal coordinate axes reflect the real relative distance information between them and the radar.

After performing first-order motion compensation with compensation phase term built by the trajectory defined in Figure 4.3a, all the four targets are registered to the "correct" positions in the nominal coordinate axes. Though target 2 and 3 are now focused, by comparing to Figure 4.3b, target 1 and 4 are deteriorated by the artificial motion errors.

Figure 4.3d shows the SAR image generated after the octave division motion compensation algorithm. According to the division principle introduced in subsection 4.1.1, the whole scene is divided into 5 segment and processed individually, then stitched together. Comparing with Figure 4.3b and 4.3c, all the 4 targets in Figure 4.3d are better

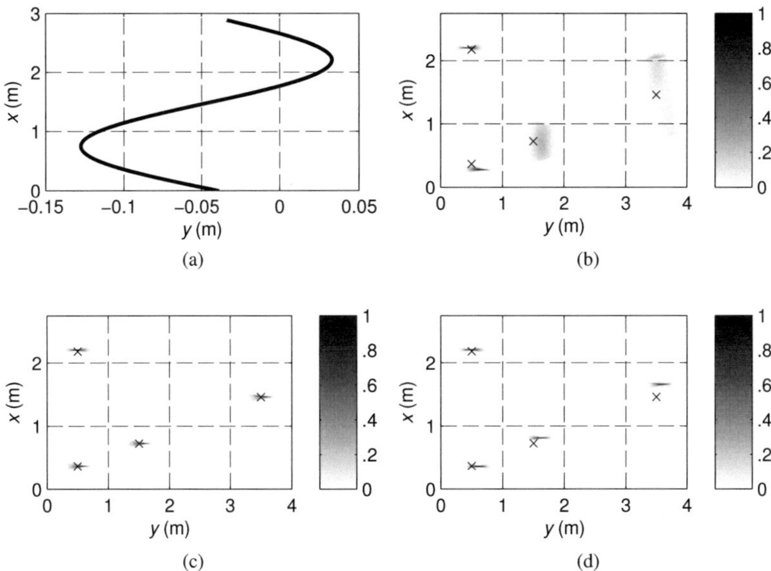

Figure 4.3: Comparison of normal first-order motion compensation and octave division algorithm with 4 point targets labeled with "x" located at their nominal
positions in the cases of $f_c = 24\,\text{GHz}$, $\Psi = 40°$: (a) applied motion error with respect to the nominal trajectory; (b) SAR image without motion
compensation; (c) SAR image with normal first-order motion compensation; (d) SAR image with octave division algorithm.

focused and their positions correctly reflect the real relative distance information with
the radar.

4.2 Slant Range Measuring Algorithms

In this work only the hardware-based motion measuring approach is considered. Generally, motion errors are measured by inertial measurement units (IMU) which are
mounted directly onto the vehicle-radar rigid body to form a strapdown system. The

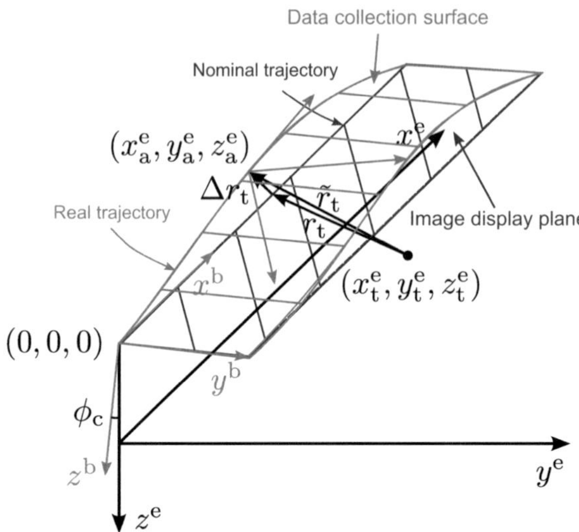

Figure 4.4: Geometry of SAR imaging and strapdown inertial navigation.

accelerometers measure the accelerations in the 3 axes of the body frame and the gyroscopes measure the angular rates about the 3 axes of the body frame. The coordinate frames referred to during the course of the following discussion are orthogonal, right-handed axis sets in which positive rotations about each axis are taken to be in a clockwise direction looking along the axis from the origin.

As discussed in Chapter 3, the SAR processing is performed in the Earth-fixed frame, while the antenna's coordinates in the Earth-fixed frame are calculated from measurements acquired by IMU in the body frame. Extending the case illustrated in Figure 3.1, a more general geometry of SAR imaging and strapdown inertial navigation is illustrated in Figure 4.4.

The sensitive axes of the accelerometers are at right angles to one another and aligned with the body axes of the vehicle. The superscript "e" denotes the measurements expressed in the Earth-fixed frame, while the superscript "b" denotes the measurements expressed in the body frame. The image display plane (IDP) expressed in the Earth-fixed frame is specified by a constant roll angle ϕ_c between $x^e y^e$-plane and IDP, which

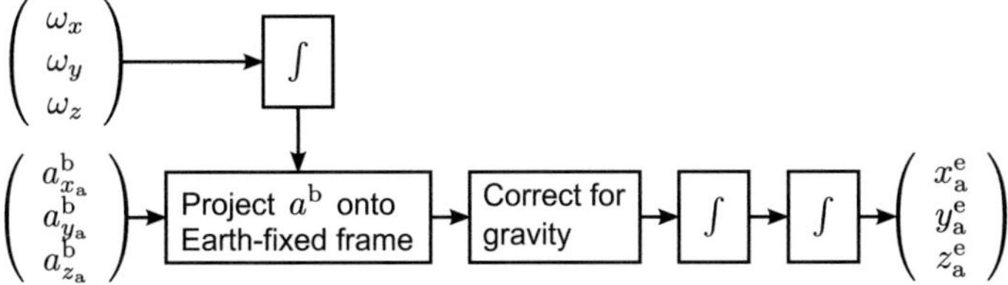

Figure 4.5: Procedure of strapdown inertial navigation algorithm.

is determined by the configuration of the radar sensor. Data collection surface (DCS) is the instantaneous $x^b y^b$-plane expressed in the Earth-fixed frame.

The traditional strapdown inertial navigation algorithm is shown in Figure 4.5.

For the near-range applications, where the navigation period is short, the effects of the rotation of the Earth on the attitude computation process can be ignored, and the Coriolis correction is no longer essential in the conversion from body frame to Earth-fixed frame [86]. Therefore, the accelerations measured in a body frame are transformed to an Earth-fixed frame using the following equation:

$$\mathbf{a}^e = \mathbf{C}_b^e \mathbf{a}^b + \mathbf{g}^e$$
$$\Longrightarrow \begin{pmatrix} a_{x_a}^e \\ a_{y_a}^e \\ a_{z_a}^e \end{pmatrix} = \begin{pmatrix} c_{11} & c_{12} & c_{13} \\ c_{21} & c_{22} & c_{23} \\ c_{31} & c_{32} & c_{33} \end{pmatrix} \begin{pmatrix} a_{x_a}^b \\ a_{y_a}^b \\ a_{z_a}^b \end{pmatrix} + \begin{pmatrix} 0 \\ 0 \\ g \end{pmatrix}, \tag{4.8}$$

where the direction cosine matrix, $\mathbf{C}_b^e$, is a 3×3 matrix which defines the attitude of the body frame with respect to the Earth-fixed frame. The element in the ith row and the jth column represents the cosine of the angle between the i-axis of the Earth-fixed frame and j-axis of the body frame. As the rotation of the Earth can be ignored for the short term navigation, $\mathbf{C}_b^e$ propagates in accordance with the following equation:

$$\dot{\mathbf{C}}_b^e = \mathbf{C}_b^e \cdot \begin{pmatrix} 0 & -\omega_z & \omega_y \\ \omega_z & 0 & -\omega_x \\ -\omega_y & \omega_x & 0 \end{pmatrix}, \tag{4.9}$$

where ω are the angular rates about the axes of the body frame measured by gyroscopes.

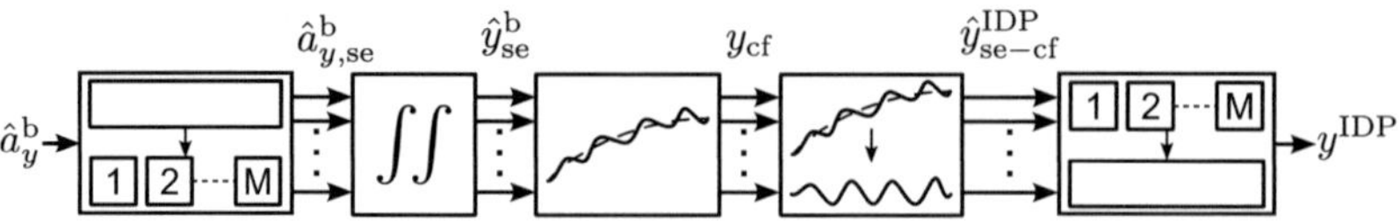

Figure 4.6: Procedure of 1-DoF motion measuring algorithm.

The acceleration vector expressed in IDP frame, $\mathbf{a}^{\mathrm{IDP}}$, can be calculated from $\mathbf{a}^{\mathrm{e}}$ as

$$
\begin{pmatrix} a_{x_a}^{\mathrm{IDP}} \\ a_{y_a}^{\mathrm{IDP}} \\ a_{z_a}^{\mathrm{IDP}} \end{pmatrix} = \begin{pmatrix} a_{x_a}^{\mathrm{e}} \\ \cos\left(\phi_c\right) a_{y_a}^{\mathrm{e}} + \sin\left(\phi_c\right) a_{z_a}^{\mathrm{e}} \\ -\sin\left(\phi_c\right) a_{y_a}^{\mathrm{e}} + \cos\left(\phi_c\right) a_{z_a}^{\mathrm{e}} \end{pmatrix}. \tag{4.10}
$$

Finally the coordinates of antenna in IDP frame can be calculated as

$$
\begin{pmatrix} x_a^{\mathrm{IDP}} \\ y_a^{\mathrm{IDP}} \\ z_a^{\mathrm{IDP}} \end{pmatrix} = \begin{pmatrix} x_{a0}^{\mathrm{IDP}} + v_{x_{a0}}^{\mathrm{IDP}} \cdot t + \iint a_{x_a}^{\mathrm{IDP}} dt^2 \\ y_{a0}^{\mathrm{IDP}} + v_{y_{a0}}^{\mathrm{IDP}} \cdot t + \iint a_{y_a}^{\mathrm{IDP}} dt^2 \\ z_{a0}^{\mathrm{IDP}} + v_{z_{a0}}^{\mathrm{IDP}} \cdot t + \iint a_{z_a}^{\mathrm{IDP}} dt^2 \end{pmatrix}. \tag{4.11}
$$

Assuming the targets are of the same height as the radar sensor, only y_a^{IDP} is used to construct the compensation phase term, $e^{jk_r \Delta r_{\mathrm{in}}}$, where $\Delta r_{\mathrm{in}} = -y_a^{\mathrm{IDP}}$.

4.2.1 1-DoF Algorithm

By assuming that rotational motions of the vehicle are minor, i.e. $\omega_x = \omega_y = \omega_z = 0$, calculation of a_y^{IDP} in (4.10) simplifies to

$$
a_y^{\mathrm{IDP}} = a_y^{\mathrm{b}} + g \cdot \sin\phi_c, \tag{4.12}
$$

where the subscript "$_a$" denoting antenna is omitted in the interest of brevity. Therefore it is possible to calculate y^{IDP} only using a one-axis accelerometer which measures a_y^{b}. The procedure of this 1-DoF algorithm for calculating y^{IDP} is shown in Figure 4.6.

Firstly, the sampled raw acceleration, $\hat{a}_y^{\mathrm{b}}$, which contains bias drift and noise is uniformly divided into M segments, the length of each segment, N_{se}, is determined by the performance of the accelerometer and the ADC. The choice of N_{se} will be discussed in detail in subsection 4.2.3. Using (4.12), the raw acceleration in each segment, $\hat{a}^b_{y,se}$, can be written as

$$
\begin{aligned}
\hat{a}^b_{y,se} &= a^b_y + a_{bias,y} + n_{a_y} \\
&= a^{IDP}_y + \left(a_{bias,y} - g \cdot \sin \phi_c\right) + n_{a_y},
\end{aligned}
\tag{4.13}
$$

where $a_{bias,y}$ is the unknown device-specific bias of the accelerometer within each segment denoted with subscript "$_{bias}$", and n_{a_y} is the overall noise of the accelerometer and ADC, which results in accumulated motion error.

After double integration, the trajectory of each segment can be obtained as

$$
\hat{y}^b_{se} = \iint \hat{a}^b_{y,se} dt = \frac{1}{2}\left(a^b_{bias,y} - g \cdot \sin \phi_c\right) t^2 + \iint a^{IDP}_y dt^2 + \iint n_{a_y} dt^2.
\tag{4.14}
$$

However, the expectation of the trajectory should be calculated as

$$
y^{IDP}_{se} = y^{IDP}_0 + v^{IDP}_{y0} \cdot t + \iint a^{IDP}_y dt^2,
\tag{4.15}
$$

where y^{IDP}_0 is the initial position and v^{IDP}_{y0} is the initial velocity of each segment.

It is seen that in order to correct errors between (4.14) and (4.15), the estimation of y^{IDP}_0, v^{IDP}_{y0} and $\frac{1}{2}\left(a^b_{bias,y} - g \cdot \sin \phi_c\right)$ are required. Performing a quadratic polynomial curve fitting on $\hat{y}^b_{se}$ is perfect for this situation, since y^{IDP}_{se} is considered to be a trajectory deviated around the vehicle's linear nominal path. The result of the curve fitting is a quadratic polynomial

$$
y_{cf} = P_1 \cdot t^2 + P_2 \cdot t + P_3.
\tag{4.16}
$$

By subtracting (4.16) from (4.14), the estimation of y^{IDP}_{se} can be obtained

$$
\hat{y}^{IDP}_{se-cf} = -P_3 - P_2 \cdot t + \iint a^{IDP}_y dt^2 + \left[\frac{1}{2}\left(a^b_{bias,y} - g \cdot \sin \phi_c\right) - P_1\right] t^2 + \iint n_{a_y} dt^2.
\tag{4.17}
$$

By comparing (4.17) with (4.15), it is seen that $-P_3$ is the estimation of y^{IDP}_0, $-P_2$ is the estimation of v^{IDP}_{y0} and P_1 contains the estimation of $\frac{1}{2}\left(a^b_{bias,y} - g \cdot \sin \phi_c\right)$. In addition, there are two advantages of performing a quadratic polynomial curve fitting: first, the device-specific $a^b_{bias,y}$ which may change along temperature and time between

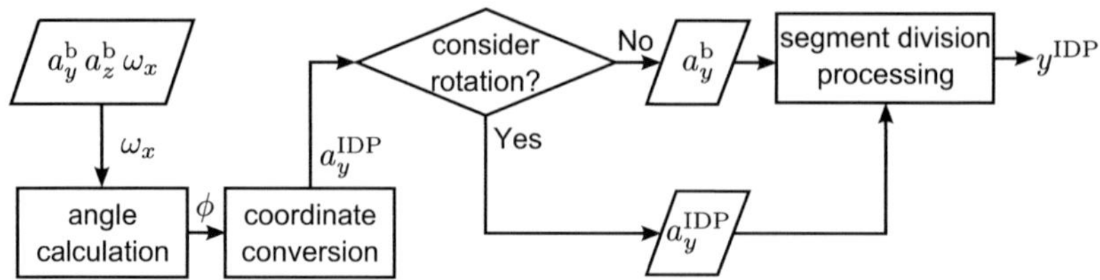

Figure 4.7: Procedure of 3-DoF motion measuring algorithm.

segments does not need to be measured; second, P_1 also compensates the quadratic component of the accumulated motion error $\iint n_{a_y} dt^2$, as will be explained in subsection 4.2.3, which considerably relaxes the requirement for the accelerometer and the ADC.

In the end, each $\hat{y}^{IDP}_{se-cf}$ is connected to its adjacent segments to form the y^{IDP} for building the motion compensation phase term, $e^{jk_r \Delta r_{in}}$.

4.2.2 3-DoF Algorithm

The rotation's influence on the accelerometer measurements cannot be ignored beyond a certain level of rotational degrees. In the 1-DoF algorithm DCS and IDP are assumed to be parallel, i.e. the change of roll angle ϕ is ignored. In order to cope with wider applications, where DCS varies over time in a three-dimensional space, a one-axis gyroscope is used to acquire ϕ information. The procedure of the 3-DoF algorithm for calculating y^{IDP} is shown in Figure 4.7.

Generally, ϕ propagates in accordance with the following differential equation:

$$\frac{d\phi}{dt} = (\omega_y \sin \phi + \omega_z \cos \phi) \tan \theta + \omega_x, \qquad (4.18)$$

where θ is the pitch angle. It can be seen that all the three angular rates are needed to calculate ϕ. However, when θ is small enough, and ω_y and ω_z are much smaller than ω_x, ϕ can be approximately calculated only by integrating ω_x over time.

Subsequently, the acceleration a_y^b and a_z^b measured by the accelerometer and ϕ calculated from the angular rate ω_x measured by the gyroscope are used to calculate the acceleration a_y^{IDP} as follows

$$a_y^{\text{IDP}} = a_y^{\text{b}}\cos(\phi - \phi_{\text{c}}) - a_z^{\text{b}}\sin(\phi - \phi_{\text{c}}) + g\sin\phi_{\text{c}}. \tag{4.19}$$

Note that (4.19) becomes (4.12) as used in the 1-DoF algorithm when ϕ is constant and equal to its initial value ϕ_{c}.

Then the decision, whether it is necessary to consider the influence of rotation, is made by calculating the ratio of the difference between rotation-including and rotation-ignoring a_y^{IDP} to rotation-ignoring a_y^{IDP} as

$$\left| \frac{a_y^{\text{b}}[\cos(\phi - \phi_{\text{c}}) - 1] - a_z^{\text{b}}\sin(\phi - \phi_{\text{c}})}{a_y^{\text{b}} + g\sin\phi_{\text{c}}} \right| \gg 1. \tag{4.20}$$

From this point, the calculation continues to the same segment division processing as in 1-DoF algorithm. In the end the complete y^{IDP} is obtained by connecting the results of each segment.

4.2.3 System and Performance Parameters

As (4.17) indicates, the performance of the motion compensation algorithm is dependent on the accumulated motion error $\iint ndt^2$ and the estimation terms: P_1, P_2 and P_3. The corresponding parameters are: the resolution of ADC, b, and the sampling frequency of the IMU's outputs, f_{IMU}; the length of each segment N_{se}; the accelerometer output bandwidth B_{acc}, bias, a_{bias}, and noise density N_{acc}; the gyroscope's bias drift, $\Delta\omega_{\text{bias}}$, and standard deviation of the noise in gyroscope's output, σ_{n_ω}.

It should be noted that though the gravity g is dependent on the latitude and altitude of the local location, the difference of gravity related to the radar's global position can be ignored comparing to the above mentioned sources of errors. For example, from the equator to the pole, g changes only 0.53%, and g at altitude of 6,000 m, which is approximately the maximum altitude of human survivability [124], is only 0.19% smaller than g at sea level.

4.2.3.1 The Choice of b and f_{IMU}

As shown in Appendix D, the standard deviation of the accumulated motion error of each segment before curve fitting, $\sigma_{\ddot{y}_{\text{se}}^{\text{b}}}$, can be expressed as a function of the bandwidth of the accelerometer, B_{acc}, and the standard deviation of the digitized acceleration, σ_{accd}, as

$$\sigma_{\ddot{y}_{\text{se}}^{\text{b}}} = \frac{\sqrt{\frac{1}{3}N_{\text{se}}^3 - \frac{5}{4}N_{\text{se}} + \frac{17}{12}N_{\text{se}} + \frac{1}{4}}}{4B_{\text{acc}}^2} \cdot \sigma_{\text{accd}}. \tag{4.21}$$

In (4.21), N_{se} is calculated from the block length L_{se} and vehicle's longitudinal velocity v as

$$N_{\text{se}} = \frac{L_{\text{se}}}{v} \cdot f_{\text{IMU}}, \tag{4.22}$$

The standard deviation of the digitized acceleration, σ_{accd}, is the sum of the standard deviation of accelerometer's noise, σ_{acc}, and the standard deviation of quantization noise introduced after digitizing acceleration by ADC, σ_{AD}.

It is common for manufacturers to specify σ_{acc} using noise density, N_{acc}. For instance, σ_{acc} of ADXL335 [125] can be calculated via N_{acc} as

$$\sigma_{\text{acc}} = N_{\text{acc}} \cdot 10^{-6} \cdot g \cdot \sqrt{1.6 \cdot B_{\text{acc}}}, \tag{4.23}$$

where B_{acc} is the bandwidth of the accelerometer and the unit of N_{acc} is $\mu g/\sqrt{\text{Hz}}$. According to [126], σ_{AD} can be calculated as a function of ADC's resolution, b, in units of bits and the accelerometer's sensitivity s_{acc}, e.g. in units of mV/g as in ADXL335 [125], which can be expressed as

$$\sigma_{\text{AD}} = \frac{1}{\sqrt{3}} \cdot \frac{V_{\text{Full}}}{2^b} \cdot \frac{10^3 \cdot g}{s_{\text{acc}}}. \tag{4.24}$$

where V_{Full} is the full voltage range of ADC, e.g. $V_{\text{Full}} = 10\,\text{V}$ for input range of $\pm 5\,\text{V}$. Given $b = 8\,\text{bits}$, $B_{\text{acc}} = 50\,\text{Hz}$ and $V_{\text{Full}} = 10\,\text{V}$, and applying the parameters of ADXL335, the ratio of σ_{AD} to σ_{acc} can be obtained as

$$\frac{\sigma_{\text{AD}}}{\sigma_{\text{acc}}} = \frac{\frac{1}{\sqrt{3}} \cdot \frac{4}{2^8} \cdot \frac{10^3 \cdot g}{300}}{150 \cdot 10^{-6} \cdot g \cdot \sqrt{1.6 \cdot 50}} \approx 56. \tag{4.25}$$

This ratio becomes 0.22 when $b = 16\,$bits. In the following calculation σ_{AD} is ignored by assuming $b = 16\,$bits.

It should be noted that (D.9) is derived under the prerequisite that $f_{\mathrm{IMU}} = 2B_{\mathrm{acc}}$, so by substituting $f_{\mathrm{IMU}} = 2B_{\mathrm{acc}}$, (4.22) and (4.23) into (4.21), the approximation of $\sigma_{\hat{y}_{\mathrm{se}}^{\mathrm{IDP}}}$ can be obtained as

$$\sigma_{\hat{y}_{\mathrm{se}}^{b}} \approx \frac{2}{\sqrt{15}} \cdot \left(\frac{L_{\mathrm{se}}}{v}\right)^{\frac{3}{2}} \cdot N_{\mathrm{acc}} \cdot 10^{-6} \cdot g. \tag{4.26}$$

It is seen that, for a certain azimuth segment L_{se}, the accumulated motion error can be reduced by driving at a higher v or choosing an accelerometer of smaller N_{acc}. Note that increasing f_{IMU} cannot improve the performance but only increases the computational cost and goes against the choice of ADC with higher resolution b. Therefore f_{IMU} should be chosen with a reasonable oversampling factor $k_{\mathrm{os}} > 1$.

4.2.3.2 The Upper Limit of N_{se}

Substituting (4.26) into (4.22), the relationship between N_{se} and $\sigma_{\hat{y}_{\mathrm{se}}^{b}}$ can be obtained as

$$N_{\mathrm{se}} = \left(\frac{\sqrt{15}}{2} \cdot \frac{\sigma_{\hat{y}_{\mathrm{se}}^{b}}}{N_{\mathrm{acc}} \cdot 10^{-6} \cdot g}\right)^{\frac{2}{3}} \cdot f_{\mathrm{IMU}}. \tag{4.27}$$

Since the major component of the accumulated motion error is quadratic as shown in Figure 4.8a, given a maximum allowable value of $\sigma_{\hat{y}_{\mathrm{se}}^{b}}$, which can be dictated according to Figure 3.23b, the upper limit of N_{se} can be determined.

In addition, as mentioned in subsection 4.2.1 the quadratic component of accumulated motion errors can be compensated via quadratic curve fitting processing. As result, the accumulated motion error in $\hat{y}_{\mathrm{se-cf}}^{\mathrm{IDP}}$ is much smaller than that in $\hat{y}_{\mathrm{se}}^{b}$, thus the upper limit of N_{se} can be improved given the same maximum allowable amplitude of the motion error. However, it is very difficult to derive a closed-form expression of $\sigma_{\hat{y}_{\mathrm{se-cf}}^{\mathrm{IDP}}}$ as a function of N_{se} and N_{acc} like (4.21). Therefore, Monte Carlo simulations have been performed, the results of 1000 times simulations are shown in Figure 4.8.

The accumulated motion errors before and after subtracting the quadratic components have been depicted in Figure 4.8a and 4.8b respectively. It is seen that $\hat{y}_{\mathrm{se}}^{b}(t)$ is quasi-quadratic while $\hat{y}_{\mathrm{se-cf}}^{\mathrm{IDP}}(t)$ is sinusoidal-like but of much smaller amplitude, and the

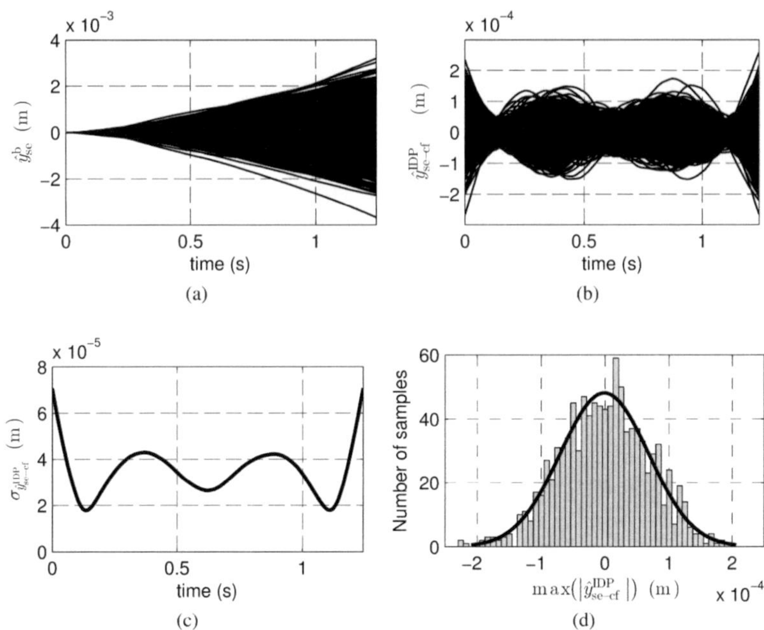

Figure 4.8: Results of Monte Carlo simulations (1000 times) of accumulated motion error: (a) $\hat{y}_{se}^{b}(t)$; (b) $\hat{y}_{se-cf}^{IDP}(t)$; (c) standard deviations of each point of $\hat{y}_{se-cf}^{IDP}$ within one segment; (d) histogram with normal fit of the maximum value of $\hat{y}_{se-cf}^{IDP}$.

frequency of this sinusoid is approximately $f_y \approx 3\frac{v}{L_{se}}$. Each point of $\hat{y}_{se-cf}^{IDP}(t)$ within the segment is normally distributed, and their standard deviations, $\sigma_{\hat{y}_{se-cf}^{IDP}}$, are shown in Figure 4.8c, which are proportional to $N_{se}^{\frac{3}{2}}$. Figure 4.8d shows the distribution of the maximum value of $\hat{y}_{se-cf}^{IDP}(t)$, which locates at the start or end of a segment as shown in Figure 4.8b and 4.8c and can be regarded as the amplitude of the accumulated sinusoidal motion errors. It is seen that the maximum value of $\sigma_{\hat{y}_{se-cf}^{IDP}}$ is only about 7% of $\sigma_{\hat{y}_{se}^{b}}$. By substituting $\sigma_{\hat{y}_{se-cf}^{IDP}} = 7\% \cdot \sigma_{\hat{y}_{se}^{b}}$ into (4.27), the relationship between N_{se} and

$\sigma_{\hat{y}_{se-cf}^{IDP}}$ can be written as

$$N_{se} \approx 9.15 \cdot \left(\frac{\sigma_{\hat{y}_{se-cf}^{IDP}}}{N_{acc} \cdot 10^{-6} \cdot g} \right)^{\frac{2}{3}} \cdot f_{IMU}. \tag{4.28}$$

Therefore, after quadratic curve fitting processing the upper limit of N_{se} should be determined by the maximum allowable $\sigma_{\hat{y}_{se-cf}^{IDP}}$ which can be regarded as the amplitude of a sinusoidal motion error of $f_y \approx 3\frac{v}{L_{se}}$.

4.2.3.3 Fundamental Constraints on v and N_{acc}

The segment division processing implies two requirements for N_{se}:

$$N_{se} > N_t = \frac{L_t}{v} \cdot f_{IMU} \tag{4.29}$$

and

$$N_{se} > k_f \cdot \frac{f_{IMU}}{f_y \,|_{MSR>0.25}}. \tag{4.30}$$

Equation (4.29) is to ensure that the constant acceleration that is ignored in each segment of length N_{se} does not contain any non-ignorable component of motion error of quadratic and sinusoidal frequency as classified by (3.6). Equation (4.30) means that for any vibration with f_y of $MSR > 0.25$, i.e. the vibration is regarded being linear to the radar, the quadratic polynomial curve fitting needs at least k_f periods of the vibration to obtain the estimations of P_1, P_2 and P_3. By substituting (3.6) into (4.30), the following relationship can be obtained

$$N_{se} > 4k_f \cdot \frac{L_t}{v} \cdot f_{IMU}. \tag{4.31}$$

As a matter of experience, k_f should be larger than 2, so the lower limit of N_{se} is determined by (4.31).

By combining (4.31) with (4.28), N_{acc} and v should meet the following requirement

$$N_{acc} \cdot 10^{-6} \cdot g < 3.46 \cdot k_f^{-\frac{3}{2}} \cdot \left(\frac{v}{L_t} \right)^{\frac{3}{2}} \cdot \sigma_{\hat{y}_{se-cf}^{IDP}}. \tag{4.32}$$

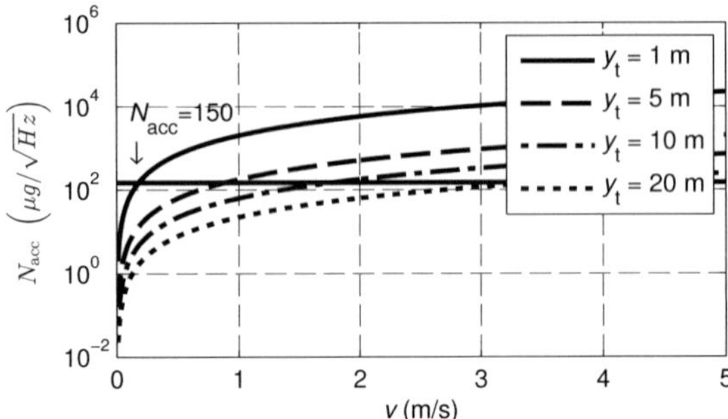

Figure 4.9: The relationship between N_{acc} and v, with $\sigma_{\hat{y}^{\text{IDP}}_{\text{se-cf}}} = 1\,\text{cm}$, $k_{\text{f}} = 2$.

According to (4.32), the requirements of N_{acc} as a function of v for different working range, y_{t}, have been calculated and depicted in Figure 4.9. For instance, assuming ADXL335 is used in the proposed motion measuring algorithm, given $N_{\text{acc}} = 150\,\mu g/\sqrt{\text{Hz}}$, the minimum velocity for the working range of $y_{\text{t}} = 15\,\text{m}$ can be obtained from Figure 4.9 as about $v_{\text{min}} = 2.7\,\text{m/s}$.

4.2.3.4 Requirement for a_{bias}

In the 1-DoF algorithm, the device-specific a_{bias} in (4.13) can be eliminated during the segment division processing. However, when (4.20) is met and the 3-DoF algorithm is used, the influence of a_{bias} has to be taken into account. Substituting a_{bias} into (4.19) yields

$$\hat{a}_y^{\text{IDP}} = a_y^b \cos\Delta\phi - a_z^b \sin\Delta\phi + g\sin\phi_{\text{c}} + \left(a_{\text{bias},y}\cos\Delta\phi - a_{\text{bias},z}\sin\Delta\phi\right), \quad (4.33)$$

where $\Delta\phi = \phi - \phi_{\text{c}}$ is the roll angle relative to the nominal one, ϕ_{c}. Since $\Delta\phi$ changes within the segment, the termed $a_{\text{bias}-y}^b \cos\Delta\phi - a_{\text{bias}-z}^b \sin\Delta\phi$ cannot be compensated as a quadratic coefficient by the quadratic curve fitting processing. In contrast, a_{bias} has to be calibrated for a specific accelerometer with enough accuracy to meet the following requirement:

$$\Delta a_{\text{bias}} \ll \sigma_{\text{acc}}, \quad (4.34)$$

where Δa_{bias} is the error remaining after calibration. Generally, at a certain temperature, $\Delta a_{\text{bias}} \approx 0$ can be achieved since a_{bias} as a function of temperature is assumed to be constant for a specific sensor [127]. Furthermore, according to the data-sheet of ADXL335 [125], for a temperature range of $10\,°C$, the change of a_{bias} is about $10\,\text{mg}$, while σ_{acc} in the case of $B_{\text{acc}} = 50\,\text{Hz}$ is $2.2\,\text{mg}$. Therefore, in high temperature dynamics conditions, an additional temperature sensor is needed for post temperature compensation to obtain acceleration measurements with low enough bias.

4.2.3.5 Requirement for $\Delta \omega_{\text{bias}}$ and σ_{ϕ_n}

By analogy with a_{bias}, the bias of a specific gyroscope, ω_{bias}, has to be accurately calibrated. However, the inevitable bias drift, $\Delta \omega_{\text{bias}}$, results in a steadily growing angular error $\Delta \omega_{\text{bias}} \cdot t$. In addition, the white noise n_ω with standard deviation σ_{n_ω} in the gyroscope's measurements results in an angle random walk $\phi_n = \int_0^t n_\omega(\tau)\,d\tau$. So the deteriorated $\hat{\phi}$ can be expressed as

$$\begin{aligned}\hat{\phi}(t) &= \int_0^t \left[\omega_x(\tau) + \Delta \omega_{\text{bias}} + n_{\omega_x}(\tau)\right] d\tau \\ &= \phi + \Delta \omega_{\text{bias}} \cdot t + \phi_n.\end{aligned} \tag{4.35}$$

As discussed in Appendix D, the standard deviation of the angle random walk, ϕ_n, can be written as

$$\sigma_{\phi_n} = \sigma_{n_w} \cdot \sqrt{\frac{t}{2B_{\text{gyr}}}} = N_{\text{gyr}} \cdot \sqrt{B_{\text{gyr}}} \cdot \sqrt{\frac{t}{2B_{\text{gyr}}}} = N_{\text{gyr}} \cdot \sqrt{\frac{t}{2}}, \tag{4.36}$$

where N_{gyr} is the noise density of the gyroscope in units of $°/s/\sqrt{\text{Hz}}$. Therefore the requirement for $\Delta \omega_{\text{bias}}$ and n_ω can be written as

$$\Delta \omega_{\text{bias}} \cdot t + \sigma_{n_\omega} \cdot \sqrt{\frac{t}{2B_{\text{gyr}}}} \ll \sigma_{\hat{\phi}}, \tag{4.37}$$

where $\sigma_{\hat{\phi}}$ is the standard deviation of $\hat{\phi}$. Taking ST's MEMS gyroscope LPR530AL [128] for example, where $N_{\text{gyr}} = 0.035\,°/s/\sqrt{\text{Hz}}$ and the bias drift for a temperature range of $10\,°C$ is $\Delta \omega_{\text{bias}} = 0.5\,°/s$, assuming $f_{\text{IMU}} = 2B_{\text{gyr}} = 100\,\text{Hz}$ and measuring for $10\,\text{s}$, the left side of (4.37) is about $5.11°$, which cannot meet the requirement dictated in (4.37). In addition, it can be seen that $\Delta\omega_{\text{bias}} \cdot t$ is much larger than $\sigma_{n_\omega} \cdot \sqrt{\frac{t}{2B_{\text{gyr}}}}$. Therefore, an additional temperature sensor is needed for post temperature compensation to reduce the bias in the gyroscope measurements.

5 SAR Demonstrator

This chapter describes a complete SAR system built for investigating the feasibility of the proposed motion measuring and compensation algorithms. The block diagram of the SAR demonstrator is depicted in Figure 5.1. The system contains an FMCW radar operating at 24 GHz [129], a rotary encoder attached on an auxiliary wheel for measuring the velocity, and a 6-DoF MEMS IMU board. The radar and motion sensors are mounted on the same board in order to be treated as a rigid body. The measurements of radar and motion sensors are synchronously sampled by two data acquisition (DAQ) modules, and the control and signal processing are performed by a PC.

Detailed descriptions of each component are given in the following sections, where the requirements of the device parameters are verified if they are not yet performed in previous chapters.

5.1 DAQ and Control Modules

The SAR system is synchronized through two DAQ modules produced by NI: USB-6251 [130] and PCI-6120 [131], where PCI-6120 works as the master device providing the clock and trigger signal to USB-6251 and the control software has been programmed in LabView. The primary parameters of them are listed in Table 5.1.

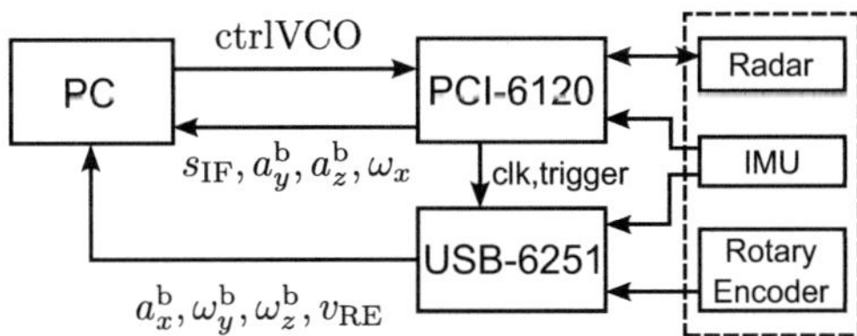

Figure 5.1: The block diagram of SAR demonstrator.

Table 5.1: Primary parameters of DAQ

Parameter	Description	PCI-6120	USB-6251
b	Input resolution	16 bits	16 bits
f_s	Maximum input sampling frequency	800 kHz	1 MHz
f_{AO}	Maximum output rate	4 MHz	2.86 MHz

The signal to control the voltage controlled oscillator (VCO) of FMCW radar is sent from PC and generated from the analog output (AO) port of PCI-6120. The VCO-controlling signal needs to be calibrated before each measurement since the VCO's characteristic line of voltage and frequency is not absolutely linear and changes slightly with time and temperature.

The IF signal s_{IF} from FMCW radar is sampled by PCI-6120 with the acceleration measurements a_y^b and a_z^b and angular rate measurements ω_x^b, while the acceleration measurements a_x^b, angular rate measurements ω_y^b and ω_z^b, and measurements of the rotary encoder v_{RE} are sampled by USB-6251. The sampling frequencies, f_s, of all these 8 channels are the same in the interest of synchronization. Therefore, the samples of IMU measurements are later decimated with a reasonable factor.

The proposed motion measuring and compensation algorithms as well as the SAR algorithm have been implemented in MATLAB. Though at current status the signal is processed in off-line mode, there will be no trouble when transplanting the code to work in real-time mode.

5.2 Radar

Though the motion compensation presented in this thesis can easily be adapted to any other radar schemes without any problems, such as pulse radar or stepped-frequency radar, the FMCW radar scheme is chosen to test and demonstrate the concept of motion compensation because of the ease in signal generation and processing. The structure and photo of the FMCW radar are shown in Figure 5.2, and the main parameters of the FMCW radar are listed in Table 5.2. The antenna's photo and beam patterns in elevation and azimuth are shown in Figure 5.3.

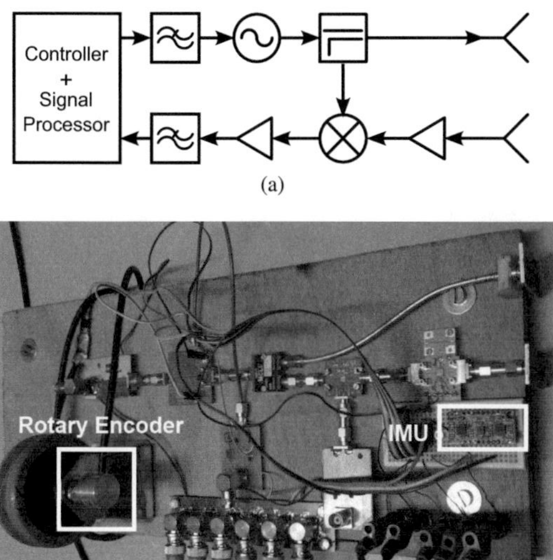

Figure 5.2: FMCW radar: (a) block diagram; (b) photo of radar and motion sensors board.

Table 5.2: FMCW radar parameters

Parameter	Description	Value
f_{min}	Minimum frequency	23.5 GHz
f_{max}	Maximum frequency	24.5 GHz
Ψ	Azimuth HPBW	33°
Φ	Elevation HPBW	17°
T_p	Duration of chirp signal	application-specific
f_s	Sampling frequency of IF signal and IMU outputs	application-specific
Δr_{ran}	Range resolution	15.0 cm
Δr_{azi}	Azimuth resolution (in SAR mode)	1.3 cm
P_t	Transmitted power of the radar system	24.6 dBm
F_{sys}	Noise figure of the radar system	48.6 dB

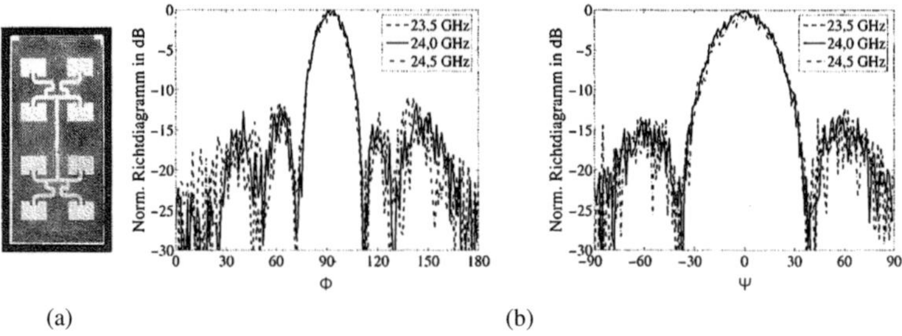

(a)   (b)

Figure 5.3: Antenna pattern: (a) photo; (b) beam patterns.

5.3 Rotary Encoder

The RM22 rotary encoder [132], which comprises a magnetic actuator and a separate encoder body, is used to obtain the radar's longitudinal position. As shown in Figure 5.2b, the rotary encoder is attached on an auxiliary wheel so the magnetic actuator can rotate with the wheel, while the encoder body senses the rotation and generate an analog signal in proportional to the rotation angle. The output of the encoder will be sampled synchronously with the radar IF signal as well as other motion measurements, so the azimuth position of every IF samples can be known.

The maximum allowable revolution per minute (*RPM*) of the rotary encoder should meet the requirement of the maximum velocity of the vehicle, the relationship can be written as

$$\frac{RPM}{60} \cdot 2\pi r_{\text{AW}} > v_{\text{max}}, \tag{5.1}$$

where r_{AW} is the radius of the auxiliary wheel. Applying $RPM = 30000$ of RM22 and $r_{\text{AW}} = 5\,\text{cm}$ into (5.1), the maximum allowable velocity for the measurement is $v_{\text{max}} \approx 157\,\text{m/s}$, which is far beyond the range of real measurements.

5.4 6-DoF IMU Board

The 6-DoF IMU board produced by SparkFun [133] has 6 analog outputs of the measurements: 3 accelerations and 3 angular rates. All analog outputs of the accelerometer

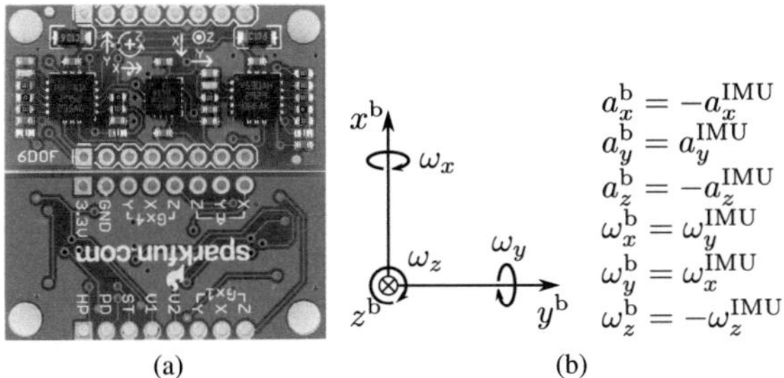

$$a_x^b = -a_x^{IMU}$$
$$a_y^b = a_y^{IMU}$$
$$a_z^b = -a_z^{IMU}$$
$$\omega_x^b = \omega_y^{IMU}$$
$$\omega_y^b = \omega_x^{IMU}$$
$$\omega_z^b = -\omega_z^{IMU}$$

Figure 5.4: 6 DoF IMU board: (a) photo; (b) original denotations with superscript "IMU" and modified denotations with superscript "b".

Table 5.3: Primary parameters of ADXL335

Parameter	Description	Value
B_{acc}	Bandwidth	$50\,\mathrm{Hz}$
s_{acc}	Sensitivity	$300\,\mathrm{mV/g}$
$N_{acc,x,y}$	Noise density in x and y axis	$150\,\mu g/\sqrt{\mathrm{Hz}}$
$N_{acc,z}$	Noise density in z axis	$300\,\mu g/\sqrt{\mathrm{Hz}}$

and gyroscopes (1x and 4x amplified) are broken out to the 2.54 mm pitch headers. The photo of the component and the original denotations of DoF and the denotations used in the SAR system are illustrated in Figure 5.4.

5.4.1 Accelerometer

The accelerometer on the IMU board is ADXL335 [125], which is a 3-axis MEMS analog accelerometer produced by Analog Devices (AD). The primary parameters of ADXL335 are listed in Table 5.3.

5.4.2 Gyroscope

Two MEMS gyroscopes are used in the IMU board: one is LPR530AL [128] for measuring ω_x and ω_y, the other one is LPR530ALH [128] for measuring ω_z. The primary

Table 5.4: Primary parameters of LPR530AL and LPR530ALH

Parameter	Description	Value
B_{gyr}	Bandwidth	$50\,\mathrm{Hz}$
s_{gyr}	Sensitivity (4x amplified)	$3.33\,\mathrm{mV}/°/\mathrm{s}$
N_{gyr}	Noise density in x and y axis	$0.035\,°/\mathrm{s}/\sqrt{\mathrm{Hz}}$
$\Delta\omega_{bias}$	Bias drift for a temperature range of $10°C$	$0.5\,°/\mathrm{s}$

parameters of these gyroscopes are listed in Table 5.4, where B_{gyr} is limited to $50\,\mathrm{Hz}$ by post low pass filtering from the nominal value of $140\,\mathrm{Hz}$.

6 Measurements

Three different vehicle-borne SAR measurements, i.e. rail-borne SAR, automotive SAR and Airquad-borne SAR, have been carried out to verify the proposed motion compensation algorithms. The rail-borne measurement is used to verify the functionality of the SAR demonstrator as the longitudinal velocity of the radar can be obtained exactly. Corresponding to the potential near-range SAR applications proposed in Chapter 1, the SAR demonstrator has been adapted and mounted on a car and an Airquad to perform imaging measurements for a typical parking lot scenario and an indoor scenario respectively. The setup and results of each measurement are given in the following sections.

6.1 Rail-borne SAR

The geometry of the rail-borne SAR measurement is denoted in Figure 6.1. The radar was mounted on a motorized sled and moved along the horizontal rail (x axis) at a constant v, while vibrations in y axis were applied by manually shaking the frame where the rail was installed. Since v was controlled via a linear positioner, it can be regarded as error-free. As shown in Figure 6.1b, a corner reflector as a point target was put at the same height level of the radar, and ADXL330 [134], the predecessor of ADXL335, of $N_{\mathrm{acc}} = 280\,\mu g/\sqrt{\mathrm{Hz}}$ was mounted beside the receiving antenna as a rigid body to measure a_y^{b}. Note that here the transmitting and receiving antennas were placed by rotating $90°$ the way as shown in Figure 5.3, so the azimuth HPBW of the rail-borne SAR was $\Psi = 17°$.

The SAR image generated with motion error-free data is shown in Figure 6.2a. It is seen that for $\Psi = 17°$ the broadening effect caused by annular support band as discussed in subsection 2.4.3 is not obvious. So the traditional definition of resolutions is used instead of the proposed definition of effective resolution. The measured range resolution is $\Delta r_{\mathrm{ran,MEFree}} = 0.91 \cdot \Delta r_{\mathrm{ran}}^{\mathrm{ideal}}$ and the measured azimuth resolution is

(a)   (b)

Figure 6.1: Setup of rail-borne SAR measurement: (a) geometry denotations, where the radar moved along the positive x axis (azimuth direction); (b) target and IMU.

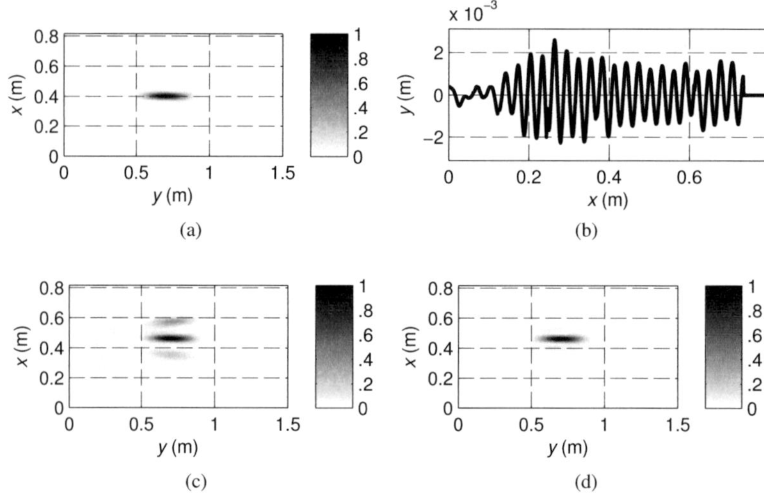

(a)   (b)

(c)   (d)

Figure 6.2: Results of rail-borne SAR measurements with $T_\mathrm{p} = 50\,\mathrm{ms}$, $f_\mathrm{s} = 4\,\mathrm{kHz}$ and $v = 0.1225\,\mathrm{m/s}$, where SAR images are normalized: (a) image without motion error; (b) measured trajectory; (c) motion-error-deteriorated image; (d) motion-error-compensated image.

$\Delta r_{\mathrm{azi,MEFree}} = 1.09 \cdot \Delta r_{\mathrm{azi}}^{\mathrm{ideal}}$, both of them meet the theoretical values quite well. In addition, the measured $ISLR_{\mathrm{MEFree}}$ is $-19.5\,\mathrm{dB}$. The subscript "$_{\mathrm{MEFree}}$" denotes the measurement corresponding to motion error-free SAR measurement.

SAR image deteriorated by sinusoidal motion errors is shown in Figure 6.2c. It is seen that sinusoidal vibration causes strong paired echoes in the SAR image. The range resolution is $\Delta r_{\mathrm{ran,ME}} = 0.92 \cdot \Delta r_{\mathrm{ran}}^{\mathrm{ideal}}$ denoted with subscript "$_{\mathrm{ME}}$", and the azimuth resolution is $\Delta r_{\mathrm{azi,MEFree}} = 1.15 \cdot \Delta r_{\mathrm{azi}}^{\mathrm{ideal}}$ which is slightly broadened. Moreover $ISLR_{\mathrm{ME}}$ degrades greatly to $-7.2\,\mathrm{dB}$ as result of the appearing of strong paired echoes.

The trajectory measured by ADXL330 using the proposed 1-DoF algorithm with $N_{\mathrm{se}} = 160$ is shown in Figure 6.2b, and the SAR image generated using the motion-error-compensated data is shown in Figure 6.2d. It is seen that most of the motion errors have been correctly measured and compensated. Denoted with subscript "$_{\mathrm{MoCo}}$", the major qualities of the SAR image are measured as $\Delta r_{\mathrm{ran,MoCo}} = 0.92 \cdot \Delta r_{\mathrm{ran}}^{\mathrm{ideal}}$, $\Delta r_{\mathrm{azi,MoCo}} = 1.08 \cdot \Delta r_{\mathrm{azi}}^{\mathrm{ideal}}$ and $ISLR_{\mathrm{MoCo}} = -21.8\,\mathrm{dB}$. It should be noted that the measurement shown in Figure 6.2c and 6.2d does not have the same starting point with the measurement shown in Figure 6.2a, so the shift in x axis between them is not caused by motion errors.

6.2 Automotive SAR

One potential near-range SAR applications is parking lot detection: the already existing automotive radar might be combined with a SAR algorithm to generate a image of the scene beside the car's trajectory with very high resolution. To verify the motion compensation algorithms in such situations, the SAR demonstrator was installed in a VW Sharan to simulate an original automotive radar.

As shown in Figure 6.3a, The radar-motion-sensor board was fixed on a wooden board and extended from the car's back door to illuminate the scene beside the car's trajectory with zero squint angle. Though v can be obtained by accessing the CAN (controller area network) bus of the car, to ease the synchronization of measurements, v was measured via rotary encoder. Note that the auxiliary wheel was not attached on the ground, instead there was a fishing line wound round the auxiliary wheel with one end stuck

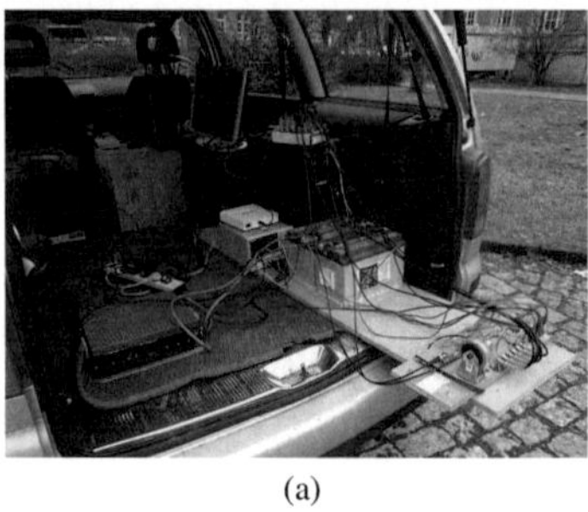

(a)   (b)

Figure 6.3: Setup of automotive SAR measurements: (a) setup of system; (b) parking lot scene with cars and empty parking spaces between them, where the car moved along the positive x axis (azimuth direction).

to a stick fixed on ground. The auxiliary wheel rotates as the car moves forward, and v was obtained using the rotary encoder's measurements.

SAR measurements were carried out in a parking lot paved with cobblestones, a typical parking scene with cars and empty parking spaces between them is shown in Figure 6.3b with denotations of the coordinate axes used in SAR imaging.

The SAR images generated without motion compensation, with motion compensation using the 1-DoF and the 3-DoF motion measuring algorithm are shown in Figure 6.4 respectively. Since the vehicle's damping system absorbed most of the vibration, the differences between the 3 SAR images in the left column are subtle, and the contours of the cars about 3 meters away from the SAR's trajectory are still recognizable even in the case without motion compensation.

Since the influence of the motion error is proportional to the target's distance, a local prominent "point scatterer" located at $(14.5\ 5.1)\,\mathrm{m}$ is chosen to demonstrate the influence of motion errors and the performance of different motion measuring algorithms. Its azimuth slices of 3 different motion measuring strategies are shown in the right column of Figure 6.4. It is seen that the defocused point target is refocused by both motion measuring algorithms while the 3-DoF algorithm has better performance than the 1-DoF algorithm. Considering the phase error caused by motion errors, $\Delta\phi\,(t)$, was less than $2°$, a higher performance improvement with the 3-DoF algorithm is expected under worse road condition, with worse vehicle's damping system, or for applications of further working range.

Note that, since the influences of the same motion error are inversely proportional to the vehicle's longitudinal velocity, motion compensation becomes much more necessary when the car drives at a lower velocity.

6.3 Airquad-borne SAR

Another potential field for near-range SAR application is the MAV based sensor system. Since a radar sensor is virtually unaffected by restrictive conditions such as weather and light quality, the SAR measurements could be used as an assistance to the current onboard optical or inferred camera system for surveillance and navigation applications.

The setup of the Airquad-borne SAR measurements is shown in Figure 6.5a. The multirotor, which is referred to as Airquad, has been supplied by the Institut für Theoretische Elektrotechnik und Systemoptimierung (ITE) of the Karlsruher Institut für Technologie (KIT). Since the whole SAR system is too heavy to be carried by the Airquad, only the radar-motion-sensor board was mounted under the landing support. The Airquad was hanged on a metal arm which was fixed on a wheel-equipped table, and the other parts of the SAR system were put on the table. The SAR measurements have been carried out by moving the table along the flying Airquad to simulate a realistic flight campaign. The inherent vibrations of the Airquad along with the maneuver have been measured by the 6-DoF IMU board. The rotary encoder measured v in an analogous way as in the automotive SAR measurements.

As shown in Figure 6.5b, the measurements were carried out in a hall. The scene consisted of several pillars, metal tables, a wall and two corner reflectors, which were approximately in the same height level with the radar's trajectory.

The results are shown in Figure 6.6. In Figure 6.6a, the pillar located at about $(3, 1)\,\text{m}$ is defocused and overlapped with the adjacent corner reflector, and the metal table located at about $(5, 3.5)\,\text{m}$ is also defocused. The Δy measured using the 1-DoF algorithm and 3-DoF algorithm are depicted in Figure 6.6b, and the compensated SAR images using these two algorithms are shown in Figure 6.6c and 6.6d respectively. It is seen

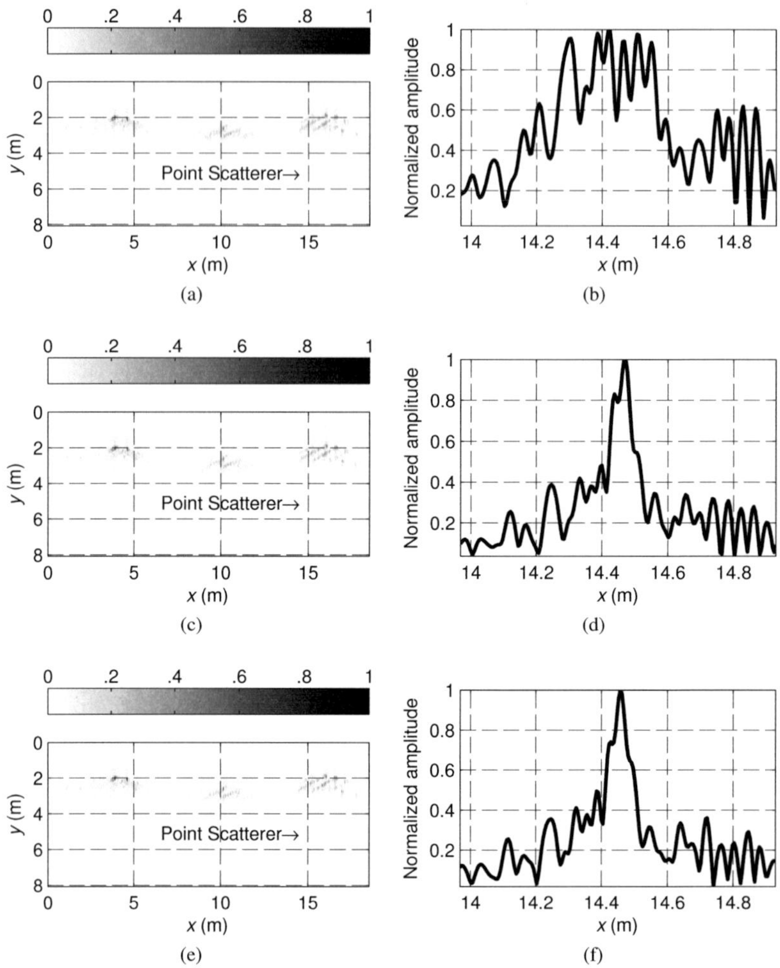

Figure 6.4: Results of automotive SAR measurements with $T_p = 2\,$ms, $f_s = 20\,$kHz, $v = 2.4\,$m/s, $f_{IMU} = 400\,$Hz and $N_{se} = 1000$, where SAR images are normalized and a local prominent "point scatterer" located at $(14.5\ 5.1)\,$m is chosen specifically: (a) and (b) image without motion compensation; (c) and (d) image compensated with 1-DoF motion measuring algorithm; (e) and (f) image compensated with 3-DoF motion measuring algorithm.

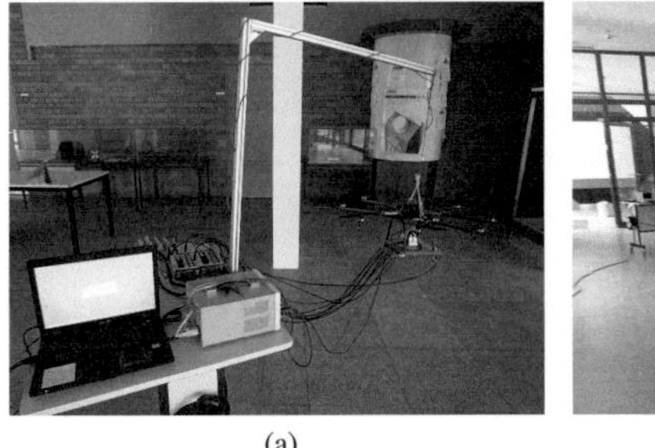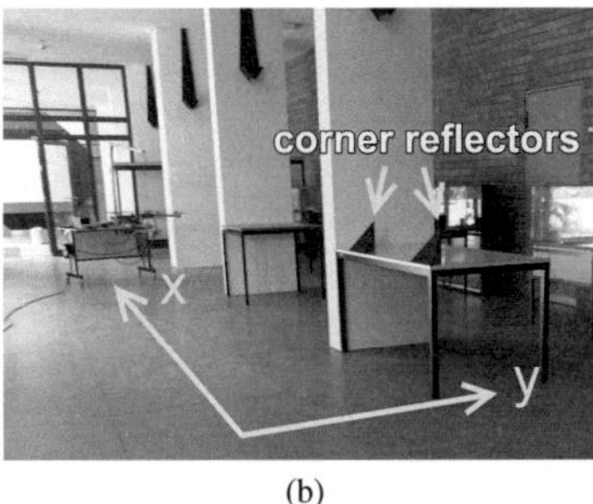

(a)　　　　　　　　　　(b)

Figure 6.5: Setup of Airquad-borne SAR measurements: (a) setup of system; (b) geometry of measurements, where the system moved along the positive x axis (azimuth direction).

that the above mentioned defocused points are refocused after motion compensation using both motion measuring algorithms. Since the motion errors deteriorated targets are not single point targets like in the rail-borne SAR or automotive SAR cases, it is difficult to give quantitative improvement comparison between the 1-DoF and 3-DoF algorithms. However, it is obvious that the 3-DoF algorithm has better performance than the 1-DoF algorithm in such high dynamics applications.

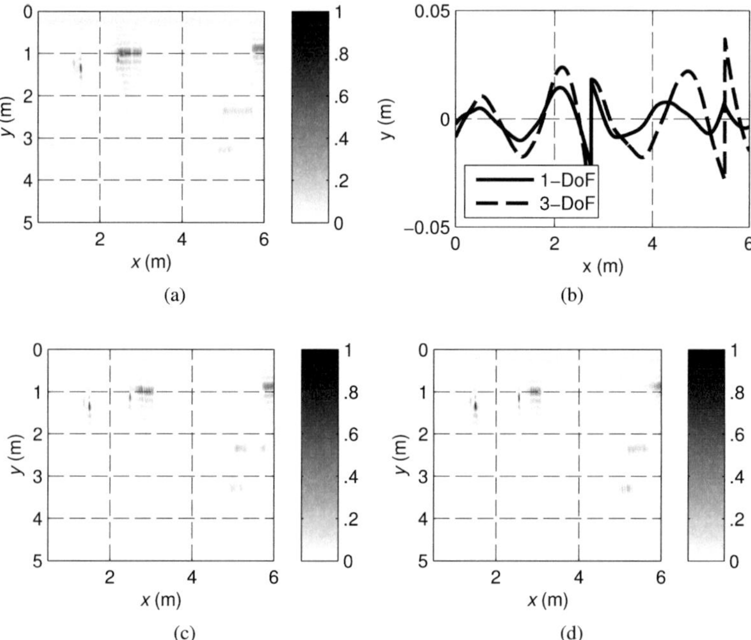

Figure 6.6: Results of Airquad-borne SAR measurements with $T_\mathrm{p} = 5\,\mathrm{ms}$, $f_\mathrm{s} = 40\,\mathrm{kHz}$, $v = 1.1\,\mathrm{m/s}$, $f_\mathrm{IMU} = 400\,\mathrm{Hz}$ and $N_\mathrm{se} = 1000$, where SAR images are normalized: (a) without motion compensation; (b) with 1-DoF motion measuring algorithm; (c) with 3-DoF motion measuring algorithm.

7 Conclusions

The concept of near-range SAR was originally proposed for the preliminary and experimental SAR studies. However, along with the democratization of radar sensor, the SAR technique combined with low-cost radar sensors has great potential for various novel applications. With their ability to work in all weather and light conditions as well as to achieve a high azimuth resolution while maintaining a wide FoV, low-cost near-range SAR systems can replace or assist some current sensor systems.

The practical use of a near-range SAR system differs in many respects from those greatly studied remote sensing SAR applications. First, the SAR algorithm's adaptability to ultra-wide antenna beamwidth was not considered, and thus great waste in terms of range and azimuth resolution could be caused if following the traditional beamwidth design guidance. Second, influences of motion errors on the SAR system with a wide beamwidth antenna was not investigated analytically, and the maximum allowable motion errors were unknown for the near-range applications. Third, the possibility of exploiting the geometry of near-range SAR for motion measurement and compensation was not explored, which could result in a substantial reduction in the requirement for hardware and computational cost.

The goal of this work is to propose a motion compensation algorithm specifically for near-range SAR applications. Corresponding to the above-mentioned research lack, the following novelties and main results have been reached and are presented in this thesis:

- *Comparison of SAR algorithms for near-range SAR applications.* A novel metric to evaluate the SAR algorithm's adaptability to a wide beamwidth has been proposed. Subsequently the optimum beamwidth for SAR systems with different operating frequency has been obtained.

- *Full investigation of the influences of motion errors.* Based on a complete dynamics model of vehicle for near-range SAR applications, the influences of motion errors have been investigated both by analytical deduction and computer simulations.

- *Octave division motion compensation algorithm.* A novel SAR motion compensation algorithm dealing with artificial motion errors specifically for near range targets has been invented. It requires less computation cost to compensate for near range targets than other compensation algorithms dealing with space-variant motion errors, while it can also be combined with some of them to compensate for far range targets.

- *1-DoF and 3-DoF motion measuring algorithms.* Two novel algorithms to measure range motion using only commercial low-cost MEMS IMU devices have been proposed for specific near-range SAR applications with quantitative analyses of requirements for the hardware.

- *First demonstration of Airquad-borne SAR.* A measurement with Airquad-borne SAR working at stripmap mode has been performed. The results were very promising. The feasibility of an operational FMCW SAR system under practical circumstances assisted by the proposed motion measuring algorithms has been proven.

A Motion Errors Design via ISO 2631

To investigate the influences of motion errors or evaluate the performances of motion compensation algorithms, realistic and complete motion parameters of the vehicle dynamics are required. One way to obtain the motion parameters is to purchase them from professional vehicle dynamics measurements company, such as Drivability Testing Alliance[135]. However, the exact measurement of the vehicle performance in driving tests is a nontrivial problem. As depicted in Figure A.1, high accuracy dynamics measurements of ground vehicle need various kinds of sensors, such as acceleration sensor, fiber optic gyroscopes, wheel vector sensor, laser sensor, GPS-based speed and position sensor, etc., thereby being expensive to carry out. Furthermore, such measurements are specific for a certain vehicle under certain circumstances instead of the whole spectrum of the vehicle dynamics, which is preferred for a guidance research.

On the other hand, in the standards ISO 2631 [114] and VDI 2057 [136], the guidelines of the effects of exposure to vibration on humans have been presented, which cover a very wide frequency bandwidth and amplitude range. Therefore it is reasonable to use such standards for designing motion parameters for the simulations with regard to evaluating the influences of motion errors and performances of the corresponding motion compensation algorithms.

In ISO 2631, the vibrations are evaluated via vibration accelerations denoted by $a(t)$, which are classified into translation vibration (in units of $\mathrm{m/s^2}$) and rotation vibration (in units of $\mathrm{rad/s^2}$). Each kind of them is measured along or about x, y and z axis separately. Considering only harmonic acceleration, which can be expressed as

$$a(t) = A_a \sin(2\pi f t), \tag{A.1}$$

the root mean square (rms) of $a(t)$ is $\frac{A_a}{\sqrt{2}}$. The frequency range under investigation is divided into a series of one-third octave bands. According to the human's perception, for each one-third octave band, a frequency weighting factor is given. For translation vibration in x and y axes, the frequency weighting factor is denoted by W_d, for

Figure A.1: Drivability testing alliance.

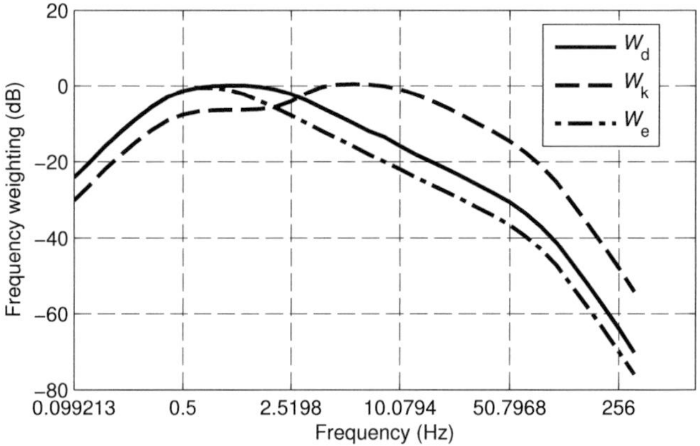

Figure A.2: Frequency weighting curves of W_d, W_k and W_e.

translation vibration in z axis the frequency weighting factor is denoted by W_k, and for rotation vibration frequency weighting factor is denoted by W_e. W_d, W_k and W_e are depicted in Figure A.2.

The overall weighted acceleration shall be determined in accordance with the following equation:

$$a_W = \sqrt{\sum_i \left(W_i \frac{A_{ai}}{\sqrt{2}} \right)^2}, \tag{A.2}$$

Table A.1: Degree of discomfort

DoD	Degree of Discomfort	a_v (m/s^2)
1	no discomfort	$(0, 0.315)$
2	a little uncomfortable	$(0.315, 0.63)$
3	fairly uncomfortable	$(0.5, 1)$
4	uncomfortable	$(0.8, 1.6)$
5	very uncomfortable	$(1.25, 2.5)$
6	extremely uncomfortable	$(2, \infty)$

where W_i is the weighting factor for the ith one-third octave band as depicted in Figure A.2, $\frac{A_{ai}}{\sqrt{2}}$ is the rms acceleration for the ith one-third octave band.

The vibration total value, a_v, of weighted rms acceleration, determined from vibration in orthogonal coordinates is calculated as follows:

$$a_v = \sqrt{k_{vx}^2 a_{Wx}^2 + k_{vy}^2 a_{Wy}^2 + k_{vz}^2 a_{Wz}^2}, \tag{A.3}$$

where a_{Wx}, a_{Wy} and a_{Wxz} are the weighted rms accelerations with respect to the orthogonal axes x, y and z respectively, and k_{vx}, k_{vy} and k_{vz} are multiplying factor. For seated persons, the multiplying factors for translation vibration are $k_{vx} = k_{vy} = k_{vz} = 1$, and the multiplying factors for rotation vibration are $k_{vx} = 0.63$ m/rad, $k_{vy} = 0.4$ m/rad and $k_{vz} = 0.2$ m/rad.

Finally, a parameter for evaluating the ride comfort called degree of discomfort (DoD) is defined via a_v. Some values giving approximate indications of likely reactions to various magnitudes of overall a_v in public transport are listed in Table A.1.

In the following, the procedure to design motion errors parameters of certain vibration, i.e. the frequency and the amplitude in (A.1), for simulations is given:

1. given y_t, z_t and Ψ, L_t is determined via (3.5);

2. given v, f_{syn} is calculated as

$$f_{syn} = \frac{v}{L_t}; \tag{A.4}$$

125

3. the frequency of investigated motion errors with respect to synthetic aperture length is calculated as $f = MSR \cdot f_{\text{syn}}$, such as

$$\text{Quadratic}: f = 0.5 f_{\text{syn}}, \tag{A.5}$$

and

$$\text{Sinusoidal}: f = f_{\text{syn}}; \tag{A.6}$$

4. corresponding amplitude of harmonic acceleration for different DoD is finally determined by

$$A_a = \frac{\sqrt{2} a_{\text{v}}}{k_{\text{v}} W(f)}. \tag{A.7}$$

B Effects of Δv_x and Δa_x

As shown in section 3.3, motion errors in azimuth direction, Δv_x and Δa_x, introduce quadratic and cubic phase errors in the spatial frequency domain. Their influences and the equivalence of their influences are discussed as follows.

For notational simplicity, the phase errors caused by Δv_x and Δa_x are expressed as the following equations:

$$\Delta\phi_{\text{azicom}}^{\Delta v_x} \approx \frac{y_t}{k_{rc}}\frac{\Delta v_x}{v}\cdot k_x^2 = K_v\cdot k_x^2, \tag{B.1}$$

$$\Delta\phi_{\text{azicom}}^{\Delta a_x} \approx \frac{y_t^2 \Delta a_x}{k_r k_{rc} v^2}\cdot k_x^3 = K_a\cdot k_x^3. \tag{B.2}$$

Then the motion errors term along k_x domain are

$$\exp\left(j\Delta\phi_{\text{azicom}}^{\Delta v_x}\right) = \exp\left(jK_v\cdot k_x^2\right), \tag{B.3}$$

and

$$\exp\left(j\Delta\phi_{\text{azicom}}^{\Delta a_x}\right) = \exp\left(jK_a\cdot k_x^3\right). \tag{B.4}$$

The inverse azimuth Fourier transforms of these motion errors terms considering the bandwidth of k_x restricted by the antenna HPBW, Ψ, are

$$\mathscr{F}^{-1}\left\{\exp\left(j\Delta\phi_{\text{azicom}}^{\Delta v_x}\right)\right\} = \int \text{rect}\left(\frac{k_x}{2k_r \sin\left(\frac{\Psi}{2}\right)}\right)\cdot \exp\left(jK_v\cdot k_x^2\right)\exp\left(jk_x x\right)dk_x, \tag{B.5}$$

and

$$\mathscr{F}^{-1}\left\{\exp\left(j\Delta\phi_{\text{azicom}}^{\Delta a_x}\right)\right\} = \int \text{rect}\left(\frac{k_x}{2k_r \sin\left(\frac{\Psi}{2}\right)}\right)\cdot \exp\left(jK_a\cdot k_x^3\right)\exp\left(jk_x x\right)dk_x, \tag{B.6}$$

where $\mathrm{rect}(\cdot)$ is a rectangular window function defined as

$$\mathrm{rect}(x) = \begin{cases} 1 & |x| < 0.5 \\ 0 & \text{others} \end{cases}. \tag{B.7}$$

POSP will be used to calculate (B.5) and (B.6) as follows.

The phase of the Fourier integrand in (B.5) is

$$\phi_{K_v}(k_x) = K_v \cdot k_x^2 + k_x x, \tag{B.8}$$

whose derivative with respect to k_x is

$$\frac{d\phi_{K_v}(k_x)}{dk_x} = 2K_v k_x + x. \tag{B.9}$$

The stationary phase can be obtained when $\frac{d\phi_{K_v}(k_x)}{dk_x} = 0$, i.e.

$$k_x = -\frac{x}{2K_v}. \tag{B.10}$$

So the result of (B.5) can be calculated as

$$\mathscr{F}^{-1}\left\{\exp\left(j\Delta\phi_{\mathrm{azicom}}^{\Delta v_x}\right)\right\} \approx \mathrm{rect}\left(\frac{-\frac{x}{2K_v}}{2k_r \sin\left(\frac{\Psi}{2}\right)}\right) \cdot \exp\left\{j\left[K_v \cdot \left(-\frac{x}{2K_v}\right)^2 + k_x x\right]\right\}$$

$$= \mathrm{rect}\left(-\frac{x}{4K_v k_r \sin\left(\frac{\Psi}{2}\right)}\right) \cdot \exp\left(-j\frac{x^2}{4K_v}\right). \tag{B.11}$$

Since a multiplication in frequency domain equals convolution in the space domain, the effect of Δv_x is broadening the IRF of SAR system by convolution with a rectangular window with length of $4K_v k_r \sin\left(\frac{\Psi}{2}\right)$.

On the other hand, the phase of the Fourier integrand in (B.6) is

$$\phi_{K_a}(k_x) = K_a \cdot k_x^3 + k_x x, \tag{B.12}$$

whose derivative with respect to k_x is

$$\frac{d\phi_{K_a}(k_x)}{dk_x} = 3K_a k_x^2 + x. \tag{B.13}$$

The points of stationary phase can be obtained when $\frac{d\phi_{K_a}(k_x)}{dk_x} = 0$, i.e.

$$k_x = \pm\sqrt{-\frac{x}{3K_a}}. \tag{B.14}$$

Since there are no real-valued solutions for (B.14) when $x \geqslant 0$, i.e. there are no stationary phase points, the stationary phase approximation of Fourier integrand is zero in the space domain where $x \geqslant 0$.

With the two points of stationary phase, (B.6) can be calculated as

$$\mathscr{F}^{-1}\left\{\exp\left(j\Delta\phi_{\text{azicom}}^{\Delta a_x}\right)\right\} \approx \text{rect}\left(\frac{\sqrt{-\frac{x}{3K_a}}}{2k_r\sin\left(\frac{\Psi}{2}\right)}\right) \cdot \exp\left\{j\sqrt{-\frac{x}{3K_a}}\left[-\frac{x}{3}+x\right]\right\}$$

$$+\text{rect}\left(\frac{-\sqrt{-\frac{x}{3K_a}}}{2k_r\sin\left(\frac{\Psi}{2}\right)}\right) \cdot \exp\left\{-j\sqrt{-\frac{x}{3K_a}}\left[-\frac{x}{3}+x\right]\right\}$$

$$= \text{rect}\left(\frac{\sqrt{-\frac{x}{3K_a}}}{2k_r\sin\left(\frac{\Psi}{2}\right)}\right) \cdot 2\cos\left(\frac{2}{3}x\sqrt{-\frac{x}{3K_a}}\right), \tag{B.15}$$

which is a real-valued function oscillating about zero within $-3K_a k_r^2 \sin^2\left(\frac{\Psi}{2}\right) \leqslant x \leqslant 0$.

Ignoring the oscillating within the space domain window, the effect of Δa_x can also be seen as a broadening of the mainlobe like the effect of Δv_x. Therefore, the Δa_x which causes the equivalent broadening effect as Δv_x can be obtained by making the window lengths of (B.11) and (B.15) equal as

$$4K_v k_r \sin\left(\frac{\Psi}{2}\right) = 3K_a k_r^2 \sin^2\left(\frac{\Psi}{2}\right). \tag{B.16}$$

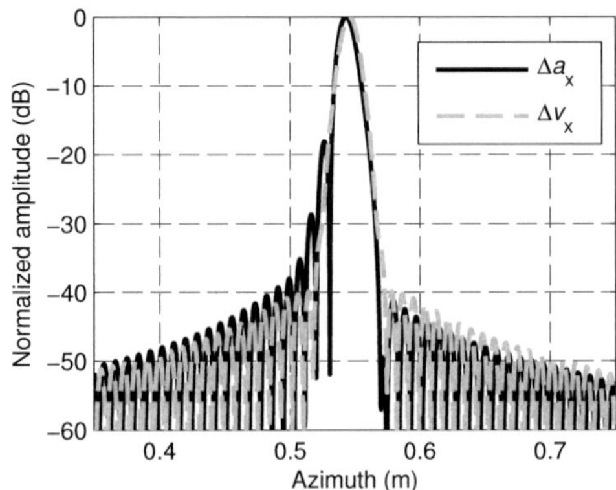

Figure B.1: Equivalence between Δa_x and Δv_x in the sense of broadening the mainlobe of system's IRF.

Substituting K_v and K_a as defined via (B.1) and (B.2) into (B.16), the equivalent Δa_x can be obtained

$$\Delta a_x = \frac{4}{3} \cdot \frac{v \Delta v_x}{y_t \sin\left(\frac{\Psi}{2}\right)}. \tag{B.17}$$

The results of (B.5) and (B.6) are depicted in Figure B.1, where Δa_x is dictated via (B.17). It can be seen that the envelopes of the inverse Fourier transform of phase errors caused by Δa_x and Δv_x are almost equal. In addition, Δa_x also causes the target to shift along azimuth direction slightly.

C Approximate Linear System for Motion Errors, Δr_{in}

Though SAR system is a linear system with respect to the motion errors for a certain target, Δr_{t}, e.g. Δr_{t} can be decomposed into Δr_{in} and Δr_{var} and compensated separately. It is not a linear system with respect to the space-variant component of motion errors, Δr_{in}. The mathematical expression is

$$\Delta r_{\text{t}} = \Delta r_{\text{in}} + \Delta r_{\text{var}} \neq \Delta r_{\text{in1}} + \Delta r_{\text{var1}} + \Delta r_{\text{in2}} + \Delta r_{\text{var2}}. \tag{C.1}$$

Assuming Δr_{in} can be decomposed into a superposition of Δr_{in1} and Δr_{in2}, the superposition of the motion errors with regard to a certain target, i.e. $\Delta r_{\text{in1}} + \Delta r_{\text{var1}} + \Delta r_{\text{in2}} + \Delta r_{\text{var2}}$, is not equal to the motion errors with regard to the same target caused by the original Δr_{in}.

It is complicated to prove (C.1) mathematically. As a contrast, an intuitive example is illustrated in Figure C.1. On the left side, a linear motion error, Δr_{in}, is decomposed into a constant deviation from the nominal trajectory, Δr_{in1}, and a linear trajectory which starts from zero, Δr_{in2}. The slant ranges at the middle of the synthetic aperture with motion errors, $\tilde{r}_{\text{t}}$, $\tilde{r}_{\text{t1}}$ and $\tilde{r}_{\text{t2}}$, and the nominal slant range, r_{t}, are depicted respectively.

Table C.1: Labels of motion errors

Label	Motion errors
a	$\tilde{r}_{\text{t}}$
b	$\tilde{r}_{\text{t1}}$
c	$\tilde{r}_{\text{t2}}$
d	r_{t}
e	Δr_{in}
f	Δr_{in1}
g	Δr_{in2}

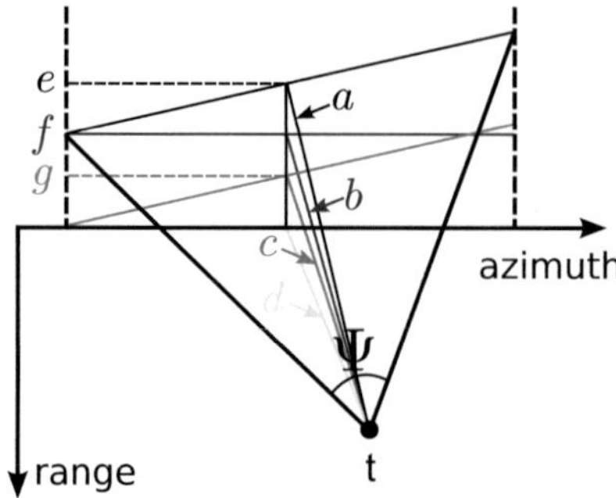

Figure C.1: Different components of motion errors.

The labels representing the length of different kinds of motion errors are listed in Table C.1. The corresponding motion errors for the target, Δr_t, Δr_{t1} and Δr_{t2}, and the space-variant components, Δr_{var}, Δr_{var1} and Δr_{var2}, can be calculated as follows:

$$\Delta r_t = a - d; \tag{C.2}$$

$$\Delta r_{t1} = b - d; \tag{C.3}$$

$$\Delta r_{t2} = c - d; \tag{C.4}$$

$$\Delta r_{\text{var}} = a - d - e; \tag{C.5}$$

$$\Delta r_{\text{var1}} = b - d - f; \tag{C.6}$$

$$\Delta r_{\text{var2}} = c - d - g. \tag{C.7}$$

Given $e = f + g$, the difference between Δr_{var} and $\Delta r_{\text{var1}} + \Delta r_{\text{var2}}$ can be obtained as

$$
\begin{aligned}
\Delta r_{\text{var}} - (\Delta r_{\text{var1}} + \Delta r_{\text{var2}}) &= a - d - e - b + d + f - c + d + g \\
&= (a + d) - (b + c) \\
&= \Delta \hat{r}_{\text{var}}. \tag{C.8}
\end{aligned}
$$

By combining (C.1) and (C.8), the motion errors of a certain target can now be expressed as

$$
\begin{aligned}
\Delta r_t &= \Delta r_{\text{in1}} + \Delta r_{\text{var1}} + \Delta r_{\text{in2}} + \Delta r_{\text{var2}} + \Delta \hat{r}_{\text{var}} \\
&= \Delta r_{t1} + \Delta r_{t2} + \Delta \hat{r}_{\text{var}}. \tag{C.9}
\end{aligned}
$$

It can be easily seen from Figure C.1 that $\Delta\hat{r}_{var}$ equals to zero only at the start of the synthetic aperture. Though it is not illustrated in Figure C.1, if Δr_{in} is decomposed into a higher constant deviation from the nominal trajectory and a linear trajectory which crosses the nominal trajectory at the middle of the linear trajectory, it is easy to deduce that $\Delta\hat{r}_{var}$ equals to zero only at the middle point of the synthetic aperture.

Two conclusions can be drawn from above discussion. First, the difference between the influences caused by Δr_{in} and the superposition of Δr_{in1} and Δr_{in2} depends on the decomposition way of Δr_{in}. Second, since the difference arises from $\Delta\hat{r}_{var}$, it is also proportional to Ψ as Δr_{var} as illustrated in Figure 3.2.

However, $\Delta\hat{r}_{var}$ can be ignored when Ψ is small, which is the implied prerequisite of literature discussing about influences of motion errors, where Δr_{in} are classified into different types according to their influences. Furthermore, even for SAR with wide Ψ as discussed in this dissertation, $\Delta\hat{r}_{var}$ is also considered to be zero, otherwise the analysis of the motion errors has to be performed for every possible combinations of different kinds of motion errors, the number of which is infinite. Therefore the SAR system is considered to be an approximate linear system for Δr_{in}, and different kinds of motion errors are analyzed separately.

D Standard Deviation of Random Walk

Considering the influences of white noise contained in the IMU measurements, the output of accelerometer and gyroscope can be expressed as

$$\hat{a} = a + n_a \tag{D.1}$$

and

$$\hat{\omega} = \omega + n_\omega, \tag{D.2}$$

where n_a and n_ω is Gauss distributed noise with zero mean and constant standard variance σ_{n_a} and σ_{n_ω} respectively.

The angle error n_ϕ resulting from integrating n_ω is a process of random walk. In this work trapezoidal integration has been used, so the accumulated error in ϕ after processing $k\,(k > 3)$ samples denoted with $n_{\phi k}$ can be expressed as

$$n_{\phi k} = \left(\frac{n_{\omega 1} + n_{\omega k}}{2} + \sum_{i=2}^{k-1} n_{\omega i} \right) \cdot \frac{1}{f_{\text{IMU}}}. \tag{D.3}$$

If $f_{\text{IMU}} = 2B_{n_\omega}$, where $B_{n_\omega} = B_{\text{gyr}}$ is the bandwidth of the gyroscope, all samples of n_ω denoted with n_{ω_i} are uncorrelated with each other and have the same standard variance σ_{n_ω}, the standard variance of $n_{\phi k}$ can be calculated as

$$\begin{aligned} \sigma_{n_\phi} &= \sigma_{n_\omega} \cdot \Delta t \cdot \sqrt{k - 1.5} \\ &\approx \sigma_{n_\omega} \cdot \sqrt{\frac{t}{2B_{\text{gyr}}}}, \end{aligned} \tag{D.4}$$

where t is the processed time span. It should be noted that increasing $f_{\text{IMU}} = k_{\text{os}} \cdot 2B_{\text{gyr}}$ with oversampling factor $k_{\text{os}} > 1$ will not decrease σ_{n_ϕ}, since the power spectrum density of n_ω and B_{gyr}, which specify the random walk of n_ϕ, are independent of f_{IMU}.

It is common for manufacturers to use angle random walk (ARW) as a noise specification in units of $°/\sqrt{h}$ or $°/\sqrt{s}$. The relationship between σ_{n_ϕ} and ARW is

$$\sigma_{n_\phi} = ARW \cdot \sqrt{t}. \tag{D.5}$$

On the other hand, by analogy with (D.3) the velocity random walk, n_{vk}, caused by n_a can be expressed as

$$n_{vk} = \left(\frac{n_{a1} + n_{ak}}{2} + \sum_{i=2}^{k-1} n_{ai} \right) \cdot \frac{1}{f_{IMU}}, \tag{D.6}$$

and the accumulated distance error (in y direction), n_{yk}, can be expressed as

$$n_{yk} = \left(\frac{n_{v1} + n_{vk}}{2} + \sum_{i=2}^{k-1} n_{vi} \right) \cdot \frac{1}{f_{IMU}}. \tag{D.7}$$

Since n_{vk} is not a white random process, the standard variance of n_{yk} cannot be calculated from $\sigma_{n_{vk}}$ directly. Instead, by substituting (D.6) into (D.7), the accumulated distance error can be expressed as a function of n_{ai} as

$$
\begin{aligned}
n_{yk} &= \left[\frac{n_{v1} \cdot f_{IMU}}{2} + \frac{1}{2} \left(\frac{n_{a_1} + n_{a_k}}{2} + \sum_{j=2}^{k-1} n_{a_j} \right) + \sum_{i=2}^{k-1} \left(\frac{n_{a_1} + n_{a_i}}{2} + \sum_{j=2}^{i-1} n_{a_j} \right) \right] \cdot \frac{1}{f_{IMU}^2} \\
&= \frac{n_{v1} \cdot f_{IMU}}{2} + \frac{1}{f_{IMU}^2} \cdot \left[\left(\frac{k}{2} - \frac{3}{4} \right) \cdot n_{a_1} + (k-2) \cdot n_{a_2} + \cdots + n_{a_{k-1}} + \frac{1}{4} n_{a_k} \right]
\end{aligned} \tag{D.8}
$$

Again, by choosing $f_{IMU} = 2B_{n_a}$, where $B_{n_a} = B_{acc}$ is the bandwidth of the accelerometer, all samples of n_a denoted with n_{a_i} are uncorrelated with each other and have the same standard variance σ_{n_a}. Therefore the standard variance of n_{yk} can be obtained as a function of σ_{n_a} and B_{acc} and expressed as

$$\sigma_{n_{yk}} = \frac{\sqrt{\frac{1}{3}k^3 - \frac{5}{4}k^2 + \frac{17}{12}k + \frac{1}{4}}}{4B_{acc}^2} \cdot \sigma_{n_a}. \tag{D.9}$$

Bibliography

[1] Richard Stevenson. Long-Distance Car Radar. *IEEE Spectrum*, 2011.

[2] H. L. Bloecher, J. Dickmann, and M. Andres. Automotive active safety & comfort functions using radar. In *Proc. IEEE Int. Conf. Ultra-Wideband ICUWB 2009*, pages 490–494, 2009.

[3] Bill Fleming. Recent Advancement in Automotive Radar Systems [Automotive Electronics]. *IEEE Vehicular Technology Magazine*, 7(1):4–9, March 2012.

[4] Smart Microwave Sensors GmbH. Micro Radar Altimeter UMRR-0A, 2011.

[5] Roke Manor Research Limited. Miniature Radar Altimeter MRA Type 2, 2012.

[6] Allistair Moses, Matthew J. Rutherford, and Kimon P. Valavanis. Radar-based detection and identification for miniature air vehicles. In *2011 IEEE International Conference on Control Applications (CCA)*, pages 933–940. IEEE, September 2011.

[7] Thomas Binzer, Michael Klar, and Volker GroB. Development of 77 GHz Radar Lens Antennas for Automotive Applications Based on Given Requirements. In *2007 2nd International ITG Conference on Antennas*, pages 205–209. IEEE, March 2007.

[8] S. Tokoro, K. Kuroda, A. Kawakubo, K. Fujita, and H. Fujinami. Electronically scanned millimeter-wave radar for pre-crash safety and adaptive cruise control system. In *IEEE IV2003 Intelligent Vehicles Symposium. Proceedings*, pages 304–309. IEEE, 2003.

[9] Valeo. Valeo Raytheon Blind Spot Detection System Receives 2007 PACE Award, 2008.

[10] Matthias Steinhauer, Hans-Oliver Ruob, Hans Irion, and Wolfgang Menzel. Millimeter-Wave-Radar Sensor Based on a Transceiver Array for Automotive Applications. *IEEE Transactions on Microwave Theory and Techniques*, 56(2):261–269, 2008.

[11] Philip E. Ross. Ford Taurus Comes With Air-Force-Grade Radar. *IEEE Spectrum*, 2010.

[12] Jürgen Hasch, Eray Topak, Raik Schnabel, Thomas Zwick, Robert Weigel, and Christian Waldschmidt. Millimeter-Wave Technology for Automotive Radar Sensors in the 77 GHz Frequency Band. *IEEE Transactions on Microwave Theory and Techniques*, 60(3):845–860, March 2012.

[13] Engin Tuncer and Benjamin Friedlander. *Classical and Modern Direction-of-Arrival Estimation*. Academic Press, 2009.

[14] Michael Klotz. *An Automotive Short Range High Resolution Pulse Radar Network*. PhD thesis, Technischen Universität Hamburg-Harburg, 2002.

[15] F. Folster. Data association and tracking for automotive radar networks. *Intelligent Transportation Systems, IEEE*, 6(4):370–377, 2005.

[16] Peter Wenig, Michael Schoor, Oliver Gunther, Bin Yang, and Robert Weigel. System Design of a 77 GHz Automotive Radar Sensor with Superresolution DOA Estimation. In *2007 International Symposium on Signals, Systems and Electronics*, pages 537–540. IEEE, July 2007.

[17] R. Feger, C. Wagner, S. Schuster, S. Scheiblhofer, H. Jager, and A. Stelzer. A 77-GHz FMCW MIMO Radar Based on an SiGe Single-Chip Transceiver. *Microwave Theory and Techniques, IEEE Transactions on*, 57(5):1020–1035, May 2009.

[18] Walter G. Carrara, Ronald M. Majewski, and Ron S. Goodman. *Spotlight Synthetic Aperture Radar: Signal Processing Algorithms*. Artech Print on Demand, 1995.

[19] Ian G. Cumming and Frank H. Wong. *Digital Processing Of Synthetic Aperture Radar Data: Algorithms And Implementation*. Artech House, 2005.

[20] A. Meta, P. Hoogeboom, and Leo P. Ligthart. Signal Processing for FMCW SAR. *Geoscience and Remote Sensing, IEEE Transactions on*, 45(11):3519–3532, 2007.

[21] ImSAR. NanoSAR - World's Smallest SAR, 2008.

[22] Evan Zaugg, Matthew Edwards, David Long, and Craig Stringham. Developments in compact high-performance synthetic aperture radar systems for use on small Unmanned Aircraft. In *2011 Aerospace Conference*, pages 1–14. IEEE, March 2011.

[23] Byung-Lae Cho, Young-Kyun Kong, Hyung-Geun Park, and Young-Soo Kim. Automobile-based SAR/InSAR system for ground experiments. *Geoscience and Remote Sensing Letters, IEEE*, 3:401–405, 2006.

[24] Evan C. Zaugg, Derek L. Hudson, and David G. Long. The BYU microSAR: A Small, Student-Built SAR for UAV Operation. In *Proc. IEEE International Conference on Geoscience and Remote Sensing Symposium IGARSS 2006*, pages 411–414, 2006.

[25] Stefan Goerner and Hermann Rohling. Parking Lot Detection with 24 GHz Radar Sensor. In *WIT 2006: 3rd International Workshop on Intelligent Transportation*, 2006.

[26] Igor E. Paromtchik and Christian Laugier. Motion generation and control for parking an autonomous vehicle. In *Proc. IEEE International Conference on Robotics and Automation*, volume 4, pages 3117–3122, 1996.

[27] K. Jiang. A sensor guided parallel parking system for nonholonomic vehicles. In *Proc. IEEE Intelligent Transportation Systems*, pages 270–275, 2000.

[28] Gizmag. A new type of parking assistant - parking space measurement from Bosch. 2006.

[29] S. Lee, D. Yoon, and A. Ghosh. Intelligent parking lot application using wireless sensor networks. In *Proc. International Symposium on Collaborative Technologies and Systems CTS 2008*, pages 48–57, 2008.

[30] Volkswagen Media Services. Volkswagen unveils Park Assist Vision, 2008.

[31] N. Kaempchen, U. Franke, and R. Ott. Stereo vision based pose estimation of parking lots using 3D vehicle models. In *Proc. IEEE Intelligent Vehicle Symposium*, volume 2, pages 459–464, 2002.

[32] Thomas Reichthalhammer and Erwin M Biebl. Specification and Design of a Fast-Tuning Stepped-Frequency Synthesizer for a Short-Range-SAR-System in the K-Band. In *Proc. German Microwave Conference*, 2010.

[33] Yoshio Yamaguchi, Masashi Mitsumoto, Masakazu Sengoku, and Takeo Abe. Synthetic aperture FM-CW radar applied to the detection of objects buried in snowpack. *Geoscience and Remote Sensing, IEEE Transactions on*, 32(1):11–18, 1994.

[34] L. Carin, N. Geng, M. McClure, J. Sichina, and L. Nguyen. Ultra-wideband synthetic aperture radar for mine field detection. In *Proc. Ultra-Wideband Short-Pulse Electromagnetics 4*, pages 433–441, 1998.

[35] Tian Jin and Zhimin Zhou. Ultrawideband Synthetic Aperture Radar Landmine Detection. *IEEE Transactions on Geoscience and Remote Sensing*, 45(11):3561–3573, 2007.

[36] Joaquim Fortuny. *Efficient Algorithms for Three-Dimensional Near-Field Synthetic Aperture Radar Imaging*. PhD thesis, University of Karslruhe, 2001.

[37] Christian Fischer. *Multistatisches Radar zur Lokalisierung von Objekten im Boden*. PhD thesis, Universität Karlsruhe, 2003.

[38] K. Gu, G. Wang, and J. Li. Migration based SAR imaging for ground penetrating radar systems. *IEE Proceedings - Radar, Sonar and Navigation*, 151(5):317, 2004.

[39] Natalie Frietsch, Justus Seibold, Philipp Crocoll, Michael Weiss, and Gert Trommer. Real Time Implementation of a Vision-Based UAV Detection and Tracking System for UAV-Navigation Aiding. In *AIAA Guidance, Navigation, and Control Conference*, Portland, 2011.

[40] Stephan Weiss, Markus W Achtelik, Simon Lynen, Margarita Chli, and Roland Siegwart. Real-time Onboard Visual-Inertial State Estimation and Self-Calibration of MAVs in Unknown Environments. In *IEEE International Confe-*

rence on Robotics and Automation (ICRA), volume 231855, 2012.

[41] Evan Ackerman. Quadrotor + Kinect = One Weird Looking Robot. *IEEE Spectrum*, 2010.

[42] Mehrdad Soumekh. *Synthetic Aperture Radar Signal Processing with MATLAB Algorithms*. Wiley-Interscience, 1999.

[43] A. Reigber, M. Jager, A. Dietzsch, R. Hansch, M. Weber, H. Przybyl, and P. Prats. A distributed approach to efficient time-domain SAR processing. In *Proc. IEEE Int. Geoscience and Remote Sensing Symp. IGARSS 2007*, pages 582–585, 2007.

[44] O. Frey, C. Magnard, M. Ruegg, and E. Meier. Focusing of Airborne Synthetic Aperture Radar Data From Highly Nonlinear Flight Tracks. *IEEE Transactions on Geoscience and Remote Sensing*, 47(6):1844–1858, June 2009.

[45] S.N. Madsen and J. Dall. Processing Of The Danish C-band SAR Data. In *10th Annual International Symposium on Geoscience and Remote Sensing*, pages 2029–2032, 1990.

[46] R.K. Raney, H. Runge, R. Bamler, I.G. Cumming, and F.H. Wong. Precision SAR processing using chirp scaling. *IEEE Transactions on Geoscience and Remote Sensing*, 32(4):786–799, July 1994.

[47] H. Runge and R. Bamler. A Novel High Precision SAR Focussing Algorithm Based On Chirp Scaling. In *IGARSS '92 International Geoscience and Remote Sensing Symposium*, pages 372–375. IEEE, 1992.

[48] G. Fornaro. Trajectory deviations in airborne SAR: analysis and compensation. 35(3):997–1009, 1999.

[49] Dominique Derauw and Christian Barbier. Chirp-Z Transform Based Interpolator Adaptation for Range Migration Correction. In *7th European Conference on Synthetic Aperture Radar*, pages 1–4, 2008.

[50] E. C. Zaugg and D. G. Long. Generalized Frequency-Domain SAR Processing. *IEEE Transactions on Geoscience and Remote Sensing*, 47(11):3761–3773, November 2009.

[51] M. Soumekh, D.A. Nobles, M.C. Wicks, and G.R.J. Genello. Signal proacessing of wide bandwidth and wide beamwidth P-3 SAR data. *IEEE Transactions on Aerospace and Electronic Systems*, 37(4):1122–1141, 2001.

[52] R. Goodman, S. Tummala, and W. Carrara. Issues in ultra-wideband, widebeam SAR image formation. In *Radar Conference, 1995., Record of the IEEE 1995 International Radar Conference, 1995., Record of the IEEE 1995 International*, 1995.

[53] James H. Mims and James L. Farrell. Synthetic Aperture Imaging with Maneuvers. *Aerospace and Electronic Systems, IEEE Transactions on*, 4(4):410–418, 1972.

[54] J. L. Farrell, J. H. Mims, and A. Sorrell. Effects of Navigation Errors in Maneuvering SAR. *Aerospace and Electronic Systems, IEEE Transactions on*, AES-9(5):758, 1973.

[55] John C. Kirk. Motion Compensation for Synthetic Aperture Radar. *Aerospace and Electronic Systems, IEEE Transactions on*, AES-11(3):338, 1975.

[56] Thomas A. Kennedy. Strapdown inertial measurement units for motion compensation for synthetic aperture radars. *Aerospace and Electronic Systems Magazine, IEEE*, 3(10):32, 1988.

[57] Stefan Buckreuss. Motion errors in an airborne synthetic aperture radar system. *European Transactions on Telecommunications*, 2(6):655–664, November 1991.

[58] David R. Kirk, R. Paul Maloney, and Mark E. Davis. Impact of platform motion on wide angle synthetic aperture radar (SAR) image quality. In *Radar Conference, 1999. The Record of the 1999 IEEE Radar Conference, 1999. The Record of the 1999 IEEE*, pages 41–46, 1999.

[59] G. Fornaro, E. Sansosti, R. Lanari, and M. Tesauro. Role of processing geometry in SAR raw data focusing. *IEEE Transactions on Aerospace and Electronic Systems*, 38(2):441–454, April 2002.

[60] G. Fornaro, G. Franceschetti, and S. Perna. Motion compensation errors: effects on the accuracy of airborne SAR images. 41(4):1338–1352, 2005.

[61] Young-Kyun Kong, Byoung-Lae Cho, and Young-Soo Kim. An experimental automobile-based SAR/InSAR. In *Proc. IEEE International Geoscience and Remote Sensing Symposium IGARSS '05*, volume 6, pages 4061–4064, 2005.

[62] T. Reichthalhammer and E. Biebl. Motion compensation of short-range, wide-beam synthetic aperture radar. *Advances in Radio Science*, 9:61–66, July 2011.

[63] Charles Beumier, Pascal Druyts, Yann Yvinec, and Marc Acheroy. Motion estimation of a hand-held mine detector. In *Proc. 2nd IEEE Benelux Signal Processing Symposium (SPS-2000)*, pages 1–4, Hilvarenbeek, 2000.

[64] P. Edenhofer and I. Alawneh. Hand-held operational demining system. In *5th International Conference on Antenna Theory and Techniques, 2005.*, pages 62–64. IEEE, 2005.

[65] D. Blacknell and S. Quegan. SAR Motion Compensation Using Autofocus. *International Journal of Remote Sensing*, 12:253–275, 1991.

[66] D. E. Wahl, P. H. Eichel, D. C. Ghiglia, and Jr. Jakowatz C.V. Phase gradient autofocus - a robust tool for high resolution SAR phase correction. *Aerospace and Electronic Systems, IEEE Transactions on*, 30(3):827, 1994.

[67] Wei Ye, Tat Soon Yeo, and Zheng Bao. Weighted least-squares estimation of phase errors for SAR/ISAR autofocus. *IEEE Transactions on Geoscience and Remote Sensing*, 37(5):2487–2494, 1999.

[68] J. C. Kirk, R. Lefevre, R Van Daalen Wetters, D. Woods, and B. Sullivan. Signal based motion compensation (SBMC). In *Radar Conference, 2000. The Record of the IEEE 2000 International Radar Conference, 2000. The Record of the IEEE 2000 International*, pages 463–468, 2000.

[69] Michael I. Duersch. BYU micro-SAR: a very small, low-power LFM-CW synthetic aperture radar. Master's thesis, Brigham Young University, 2004.

[70] Karlus A. CÂmara de Macedo, Rolf Scheiber, and Alberto Moreira. An Autofocus Approach for Residual Motion Errors With Application to Airborne Repeat-Pass SAR Interferometry. *IEEE Transactions on Geoscience and Remote Sensing*, 46(10):3151–3162, October 2008.

[71] Mengdao Xing, Xiuwei Jiang, Renbiao Wu, Feng Zhou, and Zheng Bao. Motion Compensation for UAV SAR Based on Raw Radar Data. 47(8):2870–2883, 2009.

[72] Kuang-Hung Liu and David C Munson. Fourier-domain multichannel autofocus for synthetic aperture radar. *IEEE transactions on image processing*, 20(12):3544–3552, December 2011.

[73] A. Meta, J. F. M. Lorga, J. J. M. de Wit, and P. Hoogeboom. Motion compensation for a high resolution Ka-band airborne FM-CW SAR. In *Radar Conference, 2005. EURAD 2005. European*, pages 391–394, 2005.

[74] Sang-Hong Park, Kyung-Tae Kim, and Dong-Hyun Kim. SAR motion compensation for Korean MUAV. *Synthetic Aperture Radar (APSAR), 2011 3rd International Asia-Pacific Conference on*, pages 1–4, 2011.

[75] J. F. M. Lorga, W. van Rossum, E. van Halsema, Q. P. Chu, and J. A. Mulder. The development of a SAR dedicated navigation system: from scratch to the first test flight. In *Position Location and Navigation Symposium, 2004. PLANS 2004*, pages 249–258, 2004.

[76] Thomas A. Kennedy. A technique for specifying navigation system performance requirements in SAR motion compensation applications. In *Position Location and Navigation Symposium, 1990. Record. 'The 1990's - A Decade of Excellence in the Navigation Sciences'. IEEE PLANS '90., IEEE*, pages 118–126, 1990.

[77] Peggy Decroix, Xavier Neyt, and Marc Acheroy. Trade-Off between Motion Measurement Accuracy and Autofocus Capabilities in Airborne SAR Motion Compensation. In *2006 International Radar Symposium*, pages 1–4. IEEE, May 2006.

[78] Sheng Wan and Eric Foxlin. Improved Pedestrian Navigation Based on Drift-Reduced MEMS IMU Chip. In *ION 2010 International Technical Meeting*. InterSense, 2010.

[79] Steven Nasiri, Martin Lim, and Mike Housholder. A Critical Review of the Market Status and Industry Challenges of Producing Consumer Grade MEMS

Gyroscopes. Technical report, InvenSense, 2010.

[80] Jessica E. Vascellaro. Drones are techies' new darlings, April 2012.

[81] Agostino Martinelli. Vision and IMU Data Fusion: Closed-Form Solutions for Attitude, Speed, Absolute Scale, and Bias Determination. *IEEE Transactions on Robotics*, 28(1):44–60, February 2012.

[82] SOH Madgwick and AJL Harrison. Estimation of IMU and MARG orientation using a gradient descent algorithm. *(ICORR), 2011 IEEE*, 2011.

[83] N. Michael, D. Mellinger, Q. Lindsey, and V. Kumar. The GRASP multiple micro UAV testbed. *IEEE Robotics and Automation Magazine*, 17(3):56–65, 2010.

[84] Xiaoying Kong. *Inertial Navigation System Algorithms for Low Cost IMU*. PhD thesis, 2000.

[85] Eun-Hwan Shin. Accuracy Improvement of Low Cost INS/GPS for Land Applications. Technical report, 2001.

[86] David H. Titterton and John L. Weston. *Strapdown inertial navigation technology*. Institution of Electrical Engineers, 2. ed. edition, 2004.

[87] A. K. Brown. GPS/INS uses low-cost MEMS IMU. *IEEE Aerospace and Electronic Systems Magazine*, 20(9):3–10, 2005.

[88] Oliver J. Woodman. An introduction to inertial navigation. Technical report, 2007.

[89] Vicon. UPENN GRASP LAB Achieving Remarkable Maneuvers for Autonomous Quadrotor Helicopter Flight. Technical report, Vicon, 2010.

[90] Xiaoping Yun, Eric R Bachmann, Hyatt Moore Iv, and James Calusdian. Self-Contained Position Tracking of Human Movement Using Small Inertial/Magnetic Sensor Modules. In *2007 IEEE International Conference on Robotics and Automation*, number April, pages 10–14, Roma, 2007.

[91] Eric Allen Johnson, Stacy J. Morris Bamberg, and Mark A. Minor. A state estimator for rejecting noise and tracking bias in inertial sensors. In *2008 IE-*

EE International Conference on Robotics and Automation, pages 3256–3263. IEEE, May 2008.

[92] A. Brandt and J. F. Gardner. Constrained navigation algorithms for strapdown inertial navigation systems with reduced set of sensors. In *American Control Conference, 1998. Proceedings of the 1998 American Control Conference, 1998. Proceedings of the 1998*, volume 3, 1998.

[93] G. Dissanayake, S. Sukkarieh, E. Nebot, and H. Durrant-Whyte. The aiding of a low-cost strapdown inertial measurement unit using vehicle model constraints for land vehicle applications. 17(5):731–747, 2001.

[94] Shaoshi Yan, Yueli Li, and Zhimin Zhou. Real-time motion compensation of an airborne UWB SAR. *Radar Conference (EuRAD), 2011 European*, pages 305–308, 2011.

[95] Yuanbao Huang, Zheng Bao, and Feng Zhou. A Novel Method for Along-Track Motion Compensation of the Airborne Strip-Map SAR. *Acta Electronica Sinica*, 33(3), 2005.

[96] H.S. Shin and J.T. Lim. Motion error correction of range migration algorithm for aircraft spotlight SAR imaging. *Radar, Sonar & Navigation, IET*, 2(2):79–85, 2008.

[97] Alberto Moreira and Yonghong Huang. Airborne SAR processing of highly squinted data using a chirp scaling approach with integrated motion compensation. 32(5):1029–1040, 1994.

[98] Alberto Moreira, J Mittermayer, and R Scheiber. Extended chirp scaling algorithm for air- and spaceborne SAR data processing in stripmap and ScanSAR imaging modes. 34(5):1123–1136, 1996.

[99] Evan C. Zaugg and David G. Long. Theory and Application of Motion Compensation for LFM-CW SAR. *IEEE Transactions on Geoscience and Remote Sensing*, 46(10):2990–2998, October 2008.

[100] E. Aliviazatos, Athanasios Potsis, A. Reigber, A. Moreira, and N. Uzunoglu. SAR processing with motion compensation using the extended wavenumber algorithm. In *Proc. EUSAR*, page 157, 2004.

[101] A. Reigber, E. Alivizatos, A. Potsis, and A. Moreira. Extended wavenumber-domain synthetic aperture radar focusing with integrated motion compensation. In *IEE Proceedings Radar, Sonar and Navigation*, volume 153, pages 301–310, 2006.

[102] K. A. C. DeMacedo and R. Scheiber. Precise Topography- and Aperture-Dependent Motion Compensation for Airborne SAR. *IEEE Geoscience and Remote Sensing Letters*, 2(2):172–176, April 2005.

[103] Walter G. Carrara. Motion compensation algorithm for widebeam stripmap SAR. In *Proceedings of SPIE*, volume 2487, pages 13–23. SPIE, June 1995.

[104] A. Potsis, A. Reigber, J. Mittermayer, Alberto Moreira, and N. Uzunoglou. Sub-aperture algorithm for motion compensation improvement in wide-beam SAR data processing. *Electronics Letters*, 37, 2001.

[105] Pau Prats and Andreas Reigber. Topography-dependent motion compensation for repeat-pass interferometric SAR systems. *Geoscience and Remote*, 2(2):206–210, 2005.

[106] R. Scheiber and V.M. Bothale. Interferometric multi-look techniques for SAR data. In *IEEE International Geoscience and Remote Sensing Symposium*, volume 1, pages 173–175. IEEE, 2002.

[107] Zhonghou Zheng, Xingzhao Liu, and Zhixin Zhou. Motion correction in synthetic aperture radar using subaperture techniques. In *2004 IEEE International Conference on Acoustics, Speech, and Signal Processing*, volume 2, pages ii–69–72. IEEE, 2004.

[108] Xiaoshuang Zheng, Weidong Yu, and Zaoshe Li. Motion compensation for wide beam SAR based on frequency division. *Journal of Electronics (China)*, 25(5):607–615, September 2008.

[109] Marwan Younis. *Digital Beam-Forming for High Resolution Wide Swath Real and Synthetic Aperture Radar*. PhD thesis, Universität Karlsruhe (TH), 2004.

[110] R. Bamler. A comparison of range-Doppler and wavenumber domain SAR focusing algorithms. *IEEE Transactions on Geoscience and Remote Sensing*, 30(4):706–713, July 1992.

[111] Davide D'Aria and Andrea Monti Guarnieri. High-Resolution Spaceborne SAR Focusing by SVD-Stolt. *IEEE Geoscience and Remote Sensing Letters*, 4(4):639–643, October 2007.

[112] Yan Liu, Guang-cai Sun, and Meng-dao Xing. A Modified ω-k Algorithm for Wide-field and High-resolution Spaceborne Spotlight SAR. *Journal of Electronics & Information Technology*, 33(9):2108–2113, 2011.

[113] J. M. Swiger. Resolution limits of ultra wideband synthetic aperture radar using a rectangular aperture for FFT processing. *Aerospace and Electronic Systems, IEEE Transactions on*, 30(3):935–938, 1994.

[114] ISO. Mechanical vibration and shock Evoluation of human exposure to whole-body vibration Part 1: General requirements, 1997.

[115] L. Lidicky and P. Hoogeboom. Matrix approach to modelling of SAR signals. In *Proc. IEEE International Radar Conference*, pages 865–870, 2005.

[116] G. Franceschetti, A. Iodice, S. Perna, and D. Riccio. Efficient simulation of airborne SAR raw data of extended scenes. *IEEE Transactions on Geoscience and Remote Sensing*, 44(10):2851–2860, October 2006.

[117] A. S. Khwaja, L. Ferro-Famil, and E. Pottier. SAR raw data simulation in case of motion errors. In *2008 IEEE Radar Conference*, pages 1–5. IEEE, May 2008.

[118] Evan Zaugg, Matthew Edwards, and Alex Margulis. The SlimSAR: A small, multi-frequency, Synthetic Aperture Radar for UAS operation. In *2010 IEEE Radar Conference*, pages 277–282. IEEE, 2010.

[119] P. Prats, K. A. de Macedo, A. Reigber, R. Scheiber, and J. J. Mallorqui. Comparison of Topography- and Aperture-Dependent Motion Compensation Algorithms for Airborne SAR. *Geoscience and Remote Sensing Letters, IEEE*, 4(3):349–353, 2007.

[120] D. Blacknell, A. Freeman, S. Quegan, I.A. Ward, I.P. Finley, C.J. Oliver, R.G. White, and J.W. Wood. Geometric accuracy in airborne SAR images. *IEEE Transactions on Aerospace and Electronic Systems*, 25(2):241–258, March 1989.

[121] Kim Se Young, Myung Noh Hoon, and Kang Min Jeong. Antenna Mask Design for SAR Performance Optimization. *IEEE Geoscience and Remote Sensing Letters*, 6(3):443–447, July 2009.

[122] Constantine A. Balanis. *Antenna Theory: Analysis And Design*. Wiley India Pvt. Ltd., 2nd edition, 2007.

[123] A. Reigber. Correction of residual motion errors in airborne SAR interferometry. *Electronics Letters*, 37(17):1083–1084, 2001.

[124] J. B. West. *Respiratory Physiology: The Essentials*. 6th edition, 1999.

[125] Analog Devices. ADXL335 Datasheet Rev B, 2010.

[126] Richard G. Lyons. *Understanding Digital Signal Processing*. 2004.

[127] VectorNav. Accelerometer Calibration, 2011.

[128] STMicroelectronics. LPR530AL MEMS Analog Gyroscope Datasheet, July 2009.

[129] Andres Markus. *System-Design und Verifikation eines 24 GHz FMCW-Radars*. PhD thesis, 2008.

[130] National Instruments. NI 625x Specifications, 2007.

[131] National Instruments. NI 6115/6120 Specifications, 2008.

[132] RLS. RM22 series non-contact rotary encoders, 2009.

[133] SparkFun. IMU 6DOF Razor - Ultra-Thin IMU, 2010.

[134] Analog Devices. ADXL330 Datasheet Rev A, 2007.

[135] DEWETRON elektronische Messgeräte GmbH. Driveability Testing Alliance. Technical report, 2011.

[136] Technical Division Vibration Technology. Human exposure to mechanical vibrations - Whole-body vibration, 2007.